THE IOWA WITH OPTIMAL MATH'S IAAT PREP BOOK

SCAN TO CHECK OUT MORE BOOKS FROM OPTIMAL MATH

www.optimalmath.com

No part of this book may be reproduced, distributed or transmitted in any form or by any means, including photocopying, recording or other electronic or mechanical methods, or used in any manner without the prior written permission of the publisher/copyright owner, except in the case of brief quotations embodied in reviews and certain other non-commercial uses permitted by copyright law.

Copyright © 2023 Optimal Math LLC. All rights reserved.

The Optimal Math Team

TABLE OF CONTENTS

Preface	5-6
Part I: Essential Mathematical Concepts	8
How to Set Up and Solve Proportions	9
What are Unit Rates?	10
How to Find Percent of a Number	11
How to Find the Greatest Common Factor	12
What are Exponents?	13
Negative Exponents	14
Scientific Notation (Base 10 Exponents)	15
Square Roots	16
Adding and Subtracting Negative and Positive Integers	17
Absolut Value	18
Order of Operations	19-20
Converting between Fractions, Decimals, and Percent	21-22
How to Solve Inequalities with Negative Variables	23
How to Find Mean, Median, Mode, and Range	24-26
How Changes to the Mean Affect Missing Numbers from the Data Set	27
How New Values Affect the Mean	28
How to Read a Table	29
How to Read a Histogram	30
How to Read a Pie Chart	31
How to Read a Scatter Plot	32
How to Read a Line Graph	33
How to Read a Stacked Bar Chart	34
How to Read a Double Bar Graph	35
Part I: Essential Mathematical Concepts	36
How to Write an Equation with a Variable to Match a Verbal Expression	37
How to Match a Function Table to a Verbal Equation	38
How to Solve Expressions with Substitution	39
What are the Four Quadrants of the X-Y Chart?	40
What Graphs Represent a Function and Which Do Not. What is a Function?	40
Solving an Equation Given an Ordered Pair	40
What is the Domain and Range of a Function?	40
How to Find the Area of a Circle	41
How to Find the Area of Triangle	41
How to Find the Perimeter of a Rectangle Given its Area and the Length of One Side	42
How to Find the Rule or Equation Given a Set of Ordered Pairs or X/Y Chart	43
How to Find the Y- Interception of Line	44
How to Find the X- Interception of Line	44
How to Find the Slope of a Line Given Two Points	45
What is Undefined Slope?	45-
How to Find the Equation of a Line that is Parallel to Another Line	47
How to Solve for Y Given X in an Equation	47
How to Write an Algebraic Equation Given a Word Problem	48
What are Supplementary Angles?	49
How to Multiply Two of the Same Variable Together	49
Practice Tests	50
Practice Tests #1	51-73
Practice Tests #2	74-90
Practice Tests #3	91-109
Practice Tests #4	110-126
Practice Tests #5	127-144
Practice Tests #6	146-162
Practice Tests #7	163-181
Practice Tests #8	182-198
Practice Tests #9	199-221
Practice Tests #10	222-240
Practice Tests Answers	241-251

Dear Parents,

Welcome to Optimal Math's inaugural IAAT text/workbook—a culmination of expertise, dedication, and understanding of the needs of young mathematical minds. This book is meticulously crafted, keeping your child's potential and aspirations at its core.

This workbook encompasses an expansive range of problems, specifically 600 unique ones, spread across 10 practice tests. It is essential to underscore that these aren't just any problems; they have been methodically designed to challenge and stretch your child's mathematical muscles. These problems are written in an engaging way with real-world application. Gone are the days of 40 watermelons and 60 bananas, instead your child will be interpreting real stock market charts and use cases of companies like SpaceX.

The rigor and complexity of the problems in this workbook are intentional. Optimal Math endeavors to ensure that students experience a level of challenge that surpasses that of the actual IAAT. The rationale behind this is straightforward: if a student can confidently tackle the problems in this workbook within the stipulated time, they can walk into the IAAT with a level of preparedness and confidence that increases their likelihood of attaining commendable results.

However, it's crucial to set clear expectations. This workbook is not an Algebra 1 textbook. While it presents problems and solutions, it does not delve into exhaustive step-by-step explanations for each answer. We recognize the plethora of quality online resources available that elucidate the solution methods exceptionally well. Our intention is not to replicate but to complement those resources, offering a focused practice platform for the IAAT.

Let me shed light on the expertise underpinning this workbook. Optimal Math isn't just another entity in the educational sphere; it is a hallmark of excellence and dedication. This book is the brainchild of seasoned educators who aren't merely teachers but specialists in prepping students for Algebra 1 and the IAAT. These educators have vast experience in navigating the nuanced requirements of the gifted and talented demographic. Their insights and expertise, amalgamated with the rigorous academic demands of these programs, make this workbook an unparalleled resource.

In entrusting Optimal Math with your child's IAAT preparation, you're not just choosing a workbook; you're aligning with a legacy of excellence, precision, and a deep-seated commitment to nurturing gifted minds. Thank you for your trust and belief in Optimal Math. Here's to a journey filled with challenges, growth, and unparalleled success.

Warm regards,
The Optimal Math Team

SECTION 1: PRE-ALGEBRAIC NUMBER SKILLS AND CONCEPTS (15 ITEMS)

Measures how well students understand some of the mathematical skills and concepts necessary to be prepared for a course in algebra. The first several questions assess students' skills with integer arithmetic. The remaining items measure conceptual understanding and problem-solving skills.

SECTION 2: INTERPRETING MATHEMATICAL INFORMATION (15 ITEMS)

Includes graphs and verbal/symbolic definitions of mathematical concepts followed by several questions about the material. This part assesses how well a student can learn new material presented graphically or textually.

SECTION 3: REPRESENTING RELATIONSHIPS (15 ITEMS)

The first five items present relationships between two sets of numbers in a table, and the students must find the general rule for the relationship. These rules involve all four arithmetic operations and positive and negative numbers. The remaining items measure how well students interpret and represent relationships from information presented in verbal, graphic, or symbolic form.

SECTION 4: USING SYMBOLS (15 ITEMS)

Items address common misconceptions about variables, equations, removing parentheses, consecutive integers, and variation of one number in an expression when others are held constant.

ESSENTIAL MATHEMATICAL CONCEPTS

HOW TO SET UP AND SOLVE PROPORTIONS

Proportions are about understanding the relationship between two sets of numbers. Remember that proportions are just fractions set equal to each other and solved by cross-multiplying. The most common mistake is setting up the proportion incorrectly.

It's just like sharing candies equally with friends. If 10 candies are shared among 2 friends equally, then 20 candies will be shared among how many friends?

> We would set up the proportion as $\dfrac{10}{2} = \dfrac{20}{X}$
>
> The proportion should have $\dfrac{Candies}{Friends} = \dfrac{Candies}{x}$
>
> Then cross multiply **10x=2(20) and** then simplify **x=4**

Imagine you and your best friend Alex are building identical LEGO models. You notice that for every 2 steps Alex completes, 5 minutes have passed. If there are 10 steps to finish the model, you can predict that it will take Alex 25 minutes to finish. The Proportion would look like $\dfrac{2}{5} = \dfrac{10}{X}$ This can be solved by cross multiplying 2x=5(10), so 2x=50 and then simplifying to x=10.

PRO TIP: Remember when you learned how to add fractions with different denominators or how to simplify fractions, you had to do the same operation to both the numerator and denominator of the fraction. You can also solve proportions in a similar way when you are first learning them. In $\dfrac{2}{5} = \dfrac{10}{X}$ ask yourself, what do I multiply 2 by to get to 10. (2x5=10) Since you must multiply by 5, you would then multiply the denominators of the fractions by 5 as well. Therefore 5x5=25.

WHAT ARE UNIT RATES?

Unit rates are simply how much of something you get for one of something else. For instance, miles per hour in a car, or price per pound in a grocery store. It's an extension of the division skills you've already learned. A common mistake is forgetting to write the units, or not realizing that the 'per' in 'per hour' means division.

FOR EXAMPLE: You and your family are on a road trip. You travel 240 miles in 4 hours. Your little sister keeps asking, "How fast are we going?" Instead of saying "240 miles in 4 hours," you simplify it to 60 miles in 1 hour or 60 miles per hour - that's finding a unit rate!

HOW TO FIND PERCENT OF A NUMBER

Understanding percentages is about recognizing that it's another way to express fractions or decimals. Fractions, Decimals and Percent are all just different forms of the same thing. The most common mistake is forgetting to move the decimal point two places when converting a whole number or decimal to a percent.

It's like figuring out how much candy you get if your friend shares a certain percentage of their 100 candies with you. Percent is always per 100. Just like you can only compare fractions with the same denominator, so we turn them into decimals to compare them easier, percent is like a fraction that always has a denominator of 100.

Percent is found by moving the decimal two places to the left, for example:

> **1.5 as a percent would be 150%**
> **0.75 as a percent would be 75%**
> **0.3 as a percent would be 30%**
> **0.09 as a percent would be 9%**
> **0.004 as a percent would be 0.4%**

To find a percent of a given number, multiply the given number by the decimal equivalent of the percent.

> **FOR EXAMPLE:** 75% of 200 would be 200(0.75)=150

HOW TO FIND THE GREATEST COMMON FACTOR

The Greatest Common Factor (GCF) is the largest number that divides evenly into two or more numbers. This concept builds on the multiplication and division facts your child already knows. The most common mistake is finding a common factor, but not the greatest one.

You and your brother have 24 and 36 stickers, respectively, and you want to divide them equally among your friends. The greatest number of friends you both could share with, without any sticker left, would be the greatest common factor (GCF) of 24 and 36, which is 12.

A good process for doing this is to list all of the factors of each number, and then picking the largest one they both share. For example the GCF of 24 and 36

> **24: 24x1, 12x2, 8x3, 6x4**
> **36: 36x1, 18x2, 12x3, 9x4, 6x6**

12 is the largest number both numbers share as a common factor. This would be used to simplify the fraction $\frac{24}{36}$. Both numbers would be divided by the GCF of 12 so the fraction will simplify to $\frac{2}{3}$.

WHAT ARE EXPONENTS?

Exponents represent how many times to use a number in a multiplication. Exponents are made up of the base and the power, also known as the exponent. In the exponent 9^2, 9 is the base, and 2 is the exponent. The exponent tells how many times the base should be multiplied by itself, so $9^2 = 9 \times 9 = 81$.

The common pitfall is to multiply the base by the exponent rather than using the base as a factor the number of times specified by the exponent. 9x2 and 9^2 are not the same thing.

Imagine you're playing a video game. In this game, the points you earn get doubled at every level. So if you have 2 points in the first level, in the 2nd level, you'll have $2^2 = 4$ points, in the 3rd level, you'll have $2^3 = 8$ points, and so on.

EXAMPLE: A puzzle box at your friend's birthday party asks you to calculate 2 to the power of 5, or 2^5 to open it. This means 2x2x2x2x2

So 2x2=4, (2x2)x2=8, (2x2x2)x2=16 and (2x2x2x2)x2=32

Or 2x2=4, 4x2=8, 8x2=16   16x2=32

EXAMPLE: To calculate 3 to the power of 4, it would be solved: $3^4 = 3 \times 3 \times 3 \times 3$
You would calculate 3^4 to be 81

NEGATIVE EXPONENTS

Exponents, just like integers, can also be negative. However, instead of making negative numbers, or numbers less than one, they make very small numbers, similar to the concepts you would have previously earned in fractions and decimals. Negative exponents are a continuation of the concept of exponents, but they introduce the idea of reciprocals (1 divided by the number). A common mistake is to think that a negative exponent makes the whole number negative, rather than making it a fraction. The larger the negative exponent, the smaller the value, just like in fractions, the larger the denominator, the less value each piece has.

A negative exponent means you have to divide, not multiply. So 5^{-2} is like dividing 1 by 5 two times, which is also $\frac{1}{5} \times \frac{1}{5}$ or $\frac{1}{25}$. So $5^{-2} = \frac{1}{5}$ then divided by 5 again. It could also be written as $\frac{1}{(5 \times 5)}$ because you have to divide by 5 twice which is the same as dividing it 25 times (5x5=25).

In a magic school, you learn that negative exponents make things smaller, not bigger. If a spell asks for 6^{-2} magic beans, it's asking for $\frac{1}{6^2} = \frac{1}{36}$ of a bean, not 36 beans. Now, that's a detail you don't want to mess up in your potion!

If you're asked to calculate 2 to the power of -3 it would be solved the same way. You know that negative exponents mean division, not multiplication, so you calculate 2^{-3} to be $\frac{1}{8}$. So $\frac{1}{(2 \times 2 \times 2)}$

Perhaps, your science teacher asks you to calculate the value of 4^{-3}. You remember that negative exponents mean division, so you calculate 4^{-3} to be $\frac{1}{64}$. Exponents are written as fractions so they do not have to be turned into decimals. It is much easier to work with 4^{-3} or even $\frac{1}{64}$ than its decimal equivalent 0.015625 which is what you would get using a calculator or long division to solve for $\frac{1}{4}$ = 0.25 ÷ 4 = 0.0625 ÷ 4 = 0.015625. You should be happy to learn about negative exponents ;)

SCIENTIFIC NOTATION (BASE 10 EXPONENTS)

Scientific notation is a way to express very large or very small numbers, making use of your knowledge of place value and exponents. Scientific notation is always written as a number starting in one's place, which can then include a decimal. It will then be multiplied by a base 10 exponent. Common pitfalls include moving the decimal point the wrong way or misunderstanding the meaning of negative exponents.

In science class, you learn about the sizes of cells and galaxies. These numbers can be super small or super large. To make it easier to write, you use scientific notation. So instead of writing out 300,000,000 meters (the distance light travels in one second), you can write it as 3×10^8 meters. This is because $10^8 = 10 \times 10 \times 10 \times 10 \times 10 \times 10 \times 10 \times 10 = 100,000,000$ and $3 \times 100,000,000 = 300,000,000$.

EXAMPLE 1: You're reading about the solar system, and you find out that the sun is about 5,000 times the size of Earth! You write this big number in scientific notation as 5×10^3.

EXAMPLE 2: In a science documentary, you hear that a certain bacterium is 0.0004 meters long. You write this tiny number in scientific notation as 4×10^{-4}.

SQUARE ROOTS

Square roots ask "what number times itself gives me this?" It extends the concept of multiplication into a new operation. If you think of exponents as super multiplication, you can think of square roots as super division. Common mistakes include confusing squares and square roots, or forgetting that both positive and negative numbers have square roots.

Let's say you're asked to find the number that when multiplied by itself gives 49. Fortunately, you are an expert on your multiplication facts and know that 7x7 =49. Therefore you know that the square root of 49 is 7.

Perhaps one day you will be in a trivia quiz. You're asked what number times itself equals 64. You know 8x8 is 64, so the square root of 64 is 8.

Square roots work very nicely with square numbers, for example $1^2, 2^2, 3^2, 4^2, 5^2$ and so on. You would also know them as all the "double" or "twin" multiplication facts, 1x1=1, 2x2=4, 3x3=9, 4x4=16, 5x5=25, etc. Square roots can get more complicated in higher levels of math.

FOR EXAMPLE: 20 has a square root. You know that 4^2=16 and 5^2=25, therefore the square root of 20 must be between 4 and 5 because 16<20<25.

ADDING AND SUBTRACTING NEGATIVE AND POSITIVE INTEGERS

This requires a good grasp of the number line and the idea that numbers can be less than zero. You have already learned about adding and subtracting, and this extends that concept. A common pitfall is forgetting to treat two negatives as a positive when they are combined together but must keep their negative sign.

For starters, negative numbers are numbers less than 0. So if you owe someone 3 dollars and you currently have no money, you actually have -3 dollars as the first 3 dollars you make won't actually go to you.

Imagine on a chilly morning, the temperature is -2 degrees. By the night, it drops another 4 degrees. The new temperature is -2 - 4 = -6 degrees. When subtracting a positive number from a negative number, the number gets even more negative.

Here is where it gets tricky- What do you do if you are subtracting a negative number? Did your mom ever tell you two wrongs don't make a right? She is correct, but feel free to let her know two negatives do make a positive :) If for example you currently owe someone 3 dollars you would have -3 dollars. If that person says they will take away your debt of -3 dollars, how much money do you have? You'd be back at 0, as -3 -(-3)=0. In Math we often view two negative signs as canceling each other out.

> How about -8 -(-5)=
> -8+5=-3
> You can also add a positive to a negative number, for example: -3+5=2

Adding and subtracting positive and negative numbers is best learned through a number line at first, and using the concept of absolute value. You do already have experience learning about it though in life through the concepts of temperature, money and maybe games with points.

ABSOLUTE VALUE

Absolute value is the distance a number is from zero on the number line. A common mistake is to think that the absolute value changes the number's sign rather than giving the distance from zero.

In a game of tag, you run 7 steps in the opposite direction. Regardless of the direction, you know the absolute value of -7 steps is still 7 steps. This is written as |7|=7 and |-7|=7.

If, in the same game you could run 5 steps backwards, or four steps forwards to evade the tagger, absolute value is just the total number of steps, regardless of the direction. The important thing to remember is that values inside of absolute value signs are always positive."**They can, however, be made negative by operations outside of the absolute value signs.**

During a family game night, you're playing a game where you move along a board. A card says "Move backward 3 spaces," but all that matters is the number of spaces you move, not the direction. That's like the absolute value; whether you moved -3 or 3, you still moved 3 spaces.

Which would have a larger absolute value, -19 or 10? -19 is further from 0, so it would have a larger absolute value.

How about the absolute values of -8 or 18? 18 would have a larger absolute value as it is farther from 0.

ORDER OF OPERATIONS

The order of operations **(PEMDAS/GEMDAS)** helps to add structure to complex expressions and uses all of the arithmetic operations you have learned so far. Common mistakes are to ignore the order and perform operations from left to right or to forget to treat what's inside parentheses as a unit. If order of operations was not followed, you could have two people solving the same math problem and get different answers. This can not happen in Math class!

It's like following a recipe to make your favorite dessert. You need to do things in a certain order to get the result you want. Just like you wouldn't put icing on a cake before baking it, in math, you have to do multiplication and division before addition and subtraction. Remember once an operation has been completed to remove the numbers and operations as they are used up. You certainly wouldn't want to double the salt in your cake (although maybe the sugar!)

You remember the order of operations, **"PEMDAS,"** which stands for Parentheses, Exponents, Multiplication and Division (from left to right), Addition and Subtraction (from left to right). This means that multiplication and division are simultaneous with each other, so you solve them from the left side of the expression and move to the right, just like when you read a book. Addition and subtraction are also simultaneous with each other. However, multiplication/division are always done before addition and subtraction, unless they were inside a parenthesis.

To solve 8 - 3 x 2.

8-6

2

EXAMPLE 1: Remember the order of operations (PEMDAS) and solve

$3 + 4 \times 2$

$3 + 8$
11

EXAMPLE 2: $(6-4)^2$
2^2
4

You remember to do the subtraction first because it is in a parenthesis, and then the exponent, so $(6-4)^2 = 4$.

CONVERTING BETWEEN FRACTIONS, DECIMALS, AND PERCENT

This is about recognizing that fractions, decimals, and percentages are different ways to express the same thing. If you understand these three concepts individually, then you have the basic skills to convert between them. The common pitfall is not understanding where to place the decimal when moving between decimals and percent or not knowing how to convert fractions to decimals using a calculator or long division.

During your class bake sale, your teacher says that you've sold $\frac{3}{4}$ of the cupcakes. To explain this to your classmates, you tell them you've sold 75% of the cupcakes (as a percent), or 0.75 of them (as a decimal). It's all the same, just written differently! To change $\frac{3}{4}$ to a decimal you would do $\frac{3}{4}=0.75$. Notice this is not 1.333 repeating. To change a decimal to a percent, you need to move the decimal two places to the right, so 0.75 is equivalent to 75%.

It's also like explaining how much pizza you ate at your party. You can say you ate $\frac{1}{2}$ of the pizza (fraction), or 0.5 of it (decimal), or 50% of it (percent). They all mean the same thing! Notice that this time when the decimal is changed to a percent, you still move the decimal point two place values, so 0.5 becomes 50.% The decimal is then removed as it is inferred that the decimal is always to the right of the ones place, so we write 50%=0.5=$\frac{1}{2}$

For example you notice that your battery is $\frac{5}{8}$ full. This is the same as 0.625 when written as a decimal and 62.5% when written as a percent. 5 divided into 8 groups makes 0.625 per group. This makes sense because if you have 5 things to share with 8 groups, each group will receive less than one thing. When changing the decimal to a percent, the decimal place is moved two palace values to the right so 0.625 becomes 62.5%. Note that you can still have a decimal value as a percent, it just represents a value that is less than 0.01 or less than 1%

Percents can be turned back into decimals, by removing the percent sign and moving the decimal two place values to the left so 9% would become 0.09 and 80% would become 0.8.

Decimals can be turned back into fractions, but this is often less efficient. This is done by placing the decimal over the place value said by its name. For example 9 hundredths which is the decimal 0.09 would be $\frac{9}{100}$. It is worth considering this does relate to percent because percent means per hundred or per cent (thing 100 cents in a dollar or 100 years in a century.) So $\frac{9}{100}$ or 0.09 would be 9 per 100 or 9%. You can even think of the % sign as a 1 separating two 0's to make the 100.

HOW TO SOLVE INEQUALITIES WITH NEGATIVE VARIABLES

Inequalities extend the concept of equations by introducing the "less than" and "greater than" symbols. This builds on your understanding of algebra and negative numbers. A common pitfall is to forget that when you multiply or divide both sides of an inequality by a negative number, the inequality sign must be flipped.

For example, in a game, you find that your points, $-2x$, are greater than 6. To figure out how many points you have, you solve $-2x > 6$ to find that $x < -3$. Note that when the variable x is isolated by multiplying both sides of the inequality by -2, the inequality must be flipped.

Or, in your game character's power, $-3y$, is less than or equal to 9. To find out your character's power, you solve $-3y \leq 9$ to find that $y \geq -3$. Note that when the variable y is isolated by multiplying both sides of the inequality by -3, the inequality must be flipped.

This applies whenever multiplying or dividing both sides by a negative number. If you were solving $3y>9$ then $y>3$. Note here, the inequality does not need to be flipped because you are not multiplying or dividing by a negative number.

HOW TO FIND MEAN, MEDIAN, MODE, AND RANGE

These are ways to summarize a bunch of numbers in a data set. The mean is the average, the median is the middle number (when numbers are ordered least to greatest), the mode is the number you see appearing the most often, and the range is the difference between the biggest and smallest numbers.

Mean, Median and Mode tell you what is similar about your data, and are called measures of center. Range tells you how different or how spread out your data set is, and is called a measure of variation.

Let's say your Math test scores for the year were 85, 90, 85, 95, and 100. The mean (average) is $\frac{(85+90+85+95+100)}{5} = 91$. Notice that the mean does not even have to be a number in the original data set. The mean is found by adding all the numbers together and then dividing by the number of numbers that you added. So you won't always divide by 5, if there were to take 6 tests you would then divide by six. If you only took one test, your mean score would just be that score, as any number divided by 1 is itself.

Mean can be thought of as a fair share, how much each number would have to give away or receive so that they all had the same amount. For example, if three friends have 3, 4 and 5 cookies, if they redistribute them equally, each friend will have 4 cookies $\frac{(3+4+5)}{3} = 4$.

Similarly, mean can be thought of as a balance point. If all the numbers in a data set were spread out in a number line where would the center of the line balance at. The Mean can also be vulnerable to outliers. For example if you are in 5th grade the average age of a student in your classroom should be between 10 and 11 as you are all about the same age. However, if you have a really old substitute teacher who was alive during the stone age, the mean age in your class could end up being much higher! Similar if your baby brother joined the class.

Back to your Math test scores for the year: 85, 90, 85, 95, and 100.

The median (middle number) is 90. Notice 90 is not necessarily in the middle according to when you took the test; the tests must be reordered in numerical order from least to greatest. 85, 85, 90, 95, 100. Now 90 is the median. If you took all the people in your class and put them in a line for picture day, if they just randomly stood up, you might not necessarily have the median-height person in the middle. This could result in some people not being in the picture, if a really tall person was in the front row. Although you might prefer this depending on how your hair looked that day!

If there are an even number of numbers in the middle, you have a little more work to do.

To find the median of 1,2,3,4 you would find both 2 and 3 are in the middle. In that case you would just find the mean of those two numbers and list it as the median. So, 2+3=5, and $\frac{5}{2}$ = 2.5. So 2.5 is the median of 1,2,3,4.

Yet again, back to your Math test scores for the year: 85, 90, 85, 95, and 100.
The mode (most common) is 85 because it is the number that appears the most often. Notice it is not just the biggest number, although it could be if you keep studying for your math tests :) You can also have data sets with no mode at all, or even multiple modes.

FOR EXAMPLE: 1,2,3 has no mode, while 1,1,2,3,3,4,4,5 actually has three modes, 1,3,4.

Lastly, your Math test scores for the year: 85, 90, 85, 95, and 100.
The range (difference between highest and lowest) is 100-85 = 15. Note that you might have to order the numbers least to greatest to find the range, as if you had scored 100 on your first and last test but still scored a low score of 85 on another unit test, you would still have a range (100-85) of 15.

The mean, median, mode, and range are fundamental statistics concepts, extending on understanding of arithmetic and ordering numbers. Students may mistake the mean for a sum, or they may confuse the mode and median. Revisiting addition, division, and ordering of numbers can help reinforce these concepts. I also highly recommend watching one of the greatest youtube videos ever made, search up "The Mean, Median and Mode Toads" by Animated Classroom.

HOW CHANGES TO THE MEAN AFFECT MISSING NUMBERS FROM THE DATA SET

If you add new data points to a data set, the average (mean) of the data will change too. If you know the average and most of the data, you can find a missing data point.

Let's say you go Bowling, and your game scores were 185, 190, 200, and one you forgot. The average score you know is 195. Can you find the missing score? It would be (195)(4) - (185+190+200) = 205. You knew the average of 4 games was 90, some multiplying it by 4 is the inverse of dividing by the number of numbers added. Then subtracting your known scores will leave you with the one score you are missing.

This could be useful if you know you need to keep a certain GPA (Grade Point Average) by the end of the semester for honor roll or stay eligible for your sports team. You could take the average you want to have, say 90% and multiply it by 1 more than your current score. You then subtract the known scores you have and it would tell you what you need to score on the next test. For example your score 89, 92, and 95 on your end of quarter tests. What will you need to average 90% in the class on the fourth quarter test?

$$(90)(4) = (89+92+95) + X$$
$$360 = 276 + X$$
$$84 = X$$

So as long as you score 84% or higher on the next test, you will average at least 90%!

HOW NEW VALUES AFFECT THE MEAN

When you add a new score or number to your list, your average (mean) might go up, down, or stay the same depending on that new number. Let's say your average unit test score is 90, and you score 95 in the next test. Will your average score go up, down, or stay the same? It will go up because your score was higher than the previous average.

If you start scoring more goals in your soccer games than your average goals per game, your average will go up! The opposite is also true, where if you score less goals in a game than your average of goals scored per game, your average would go down.

HOW TO READ A TABLE

A table is like a super-organized list. It arranges information in rows and columns so you can find and compare data easily. Tables are everywhere! The TV guide channel is a table that shows what shows are on at what times on what channels. The nutrition facts on food packages are also tables. Tables organize data and make it easier to read. Common mistakes include misreading the row or column headers or confusing rows with columns.

	Fish	Hamburger	Chicken	Wings
3:00 PM	$20.00	$30.00	$25.00	$18.00
4:00 PM	$0.00	$45.00	$0.00	$24.00
5:00 PM	$0.00	$45.00	$25.00	$125.00
6:00 PM	$20.00	$60.00	$0.00	$24.00
7:00 PM	$0.00	$75.00	$80.00	$12.00
8:00 PM	$20.00	$0.00	$100.00	$0.00

Using the table above a restaurant owner can analyze a large data set more easily. Trends can be spotted as well that may help make better decisions. For example no one seems to like the fish, or maybe chicken sells really well at dinner time.

HOW TO READ A HISTOGRAM

A histogram is a type of graph that shows how many times different ranges of numbers appear in your data. It is similar to a bar graph but includes ranges of numbers. Imagine a histogram of the ages of kids at a summer camp. It could show that there are more 9-12 year-olds than 3-6 year-olds at the camp. A common mistake includes confusing the height of the bars (frequency) with the values being represented.

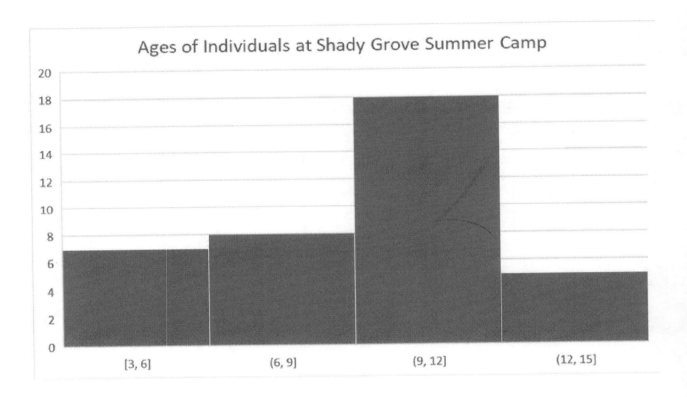

HOW TO READ A PIE CHART

A pie chart is a circle divided into parts that show what percent of the whole each part represents. Imagine a pie chart showing how you spend your time in a day. You can easily see if you spend more time sleeping or playing video games! Reading a pie chart builds on understanding fractions and percentages. Common mistakes include confusing the size of the sections with the quantity they represent.

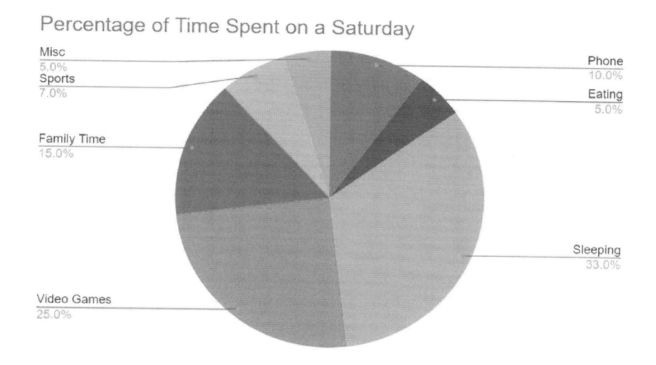

HOW TO READ A SCATTER PLOT

A scatter plot is a graph with points scattered around that show the relationship between two types of data. Imagine a scatter plot that shows the relationship between the number of hours you practice soccer and the number of goals you score. You might see that more practice usually means more goals! Or you might not, in which case you might want to talk to a coach! Scatter plots are like a game of connect-the-dots, but you don't connect the dots. They show how two things go together. A common mistake includes expecting a perfect pattern or not recognizing outlier point

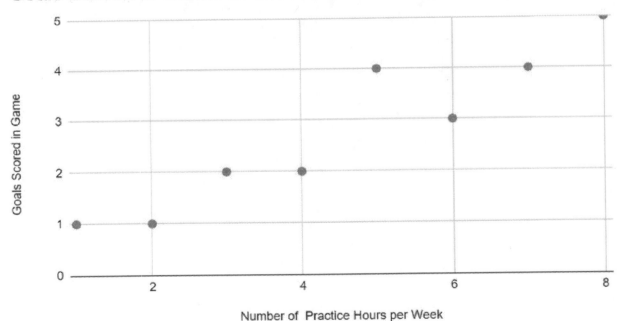

HOW TO READ A LINE GRAPH

A line graph is a way to show how something changes over time. It has a line that goes up or down as things increase or decrease. Imagine a line graph showing how your height has changed over the past year. You can see how much you've grown each month! Line graphs are typically used for things that change often, like weather temperature or stock prices. A common mistake is confusing the direction of change (up means increase, down means decrease).

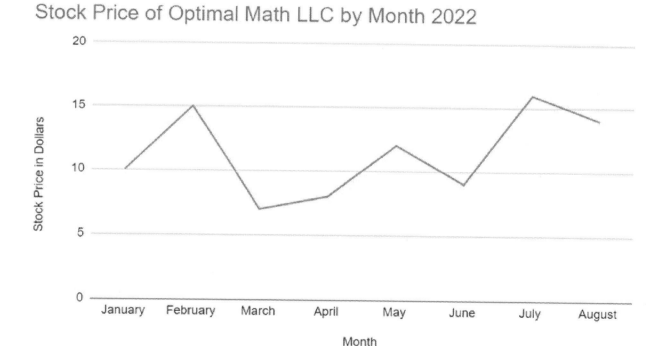

HOW TO READ A STACKED BAR CHART

A stacked bar chart shows how much each part contributes to the total, kind of like a bar made up of different colored Legos. It's like making a tower of blocks with different colors. Each color is a different part of the total tower. Imagine a stacked bar chart showing how a class's total score is made up of homework, quizzes, and tests. You can see what part each contributes to the total score. A common mistake is confusing the segments' height with their contribution to the total. Reviewing fractions and percentages can help reinforce this.

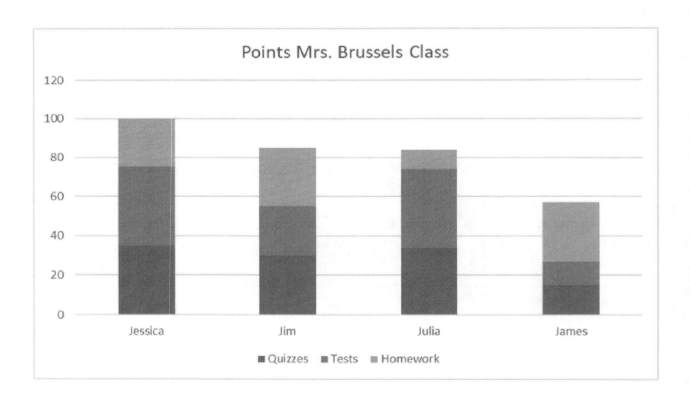

HOW TO READ A DOUBLE BAR GRAPH

A double bar graph is like a regular bar graph, but it compares two sets of data at the same time. Imagine a double bar graph comparing how many goals you and your friend scored in each soccer game. You can see who scored more goals in each game and overall. It's like having two towers of blocks for each game—one for your goals and one for your friend's. Reading a double bar graph extends from understanding bar graphs and introduces comparison of two data sets. Common mistakes include confusing which bar represents which data set or misunderstanding the scale.

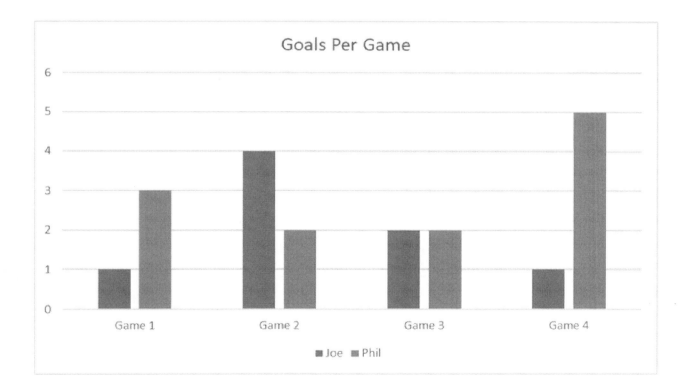

EQUATIONS AND FUNCTIONS

HOW TO WRITE AN EQUATION WITH A VARIABLE TO MATCH A VERBAL EXPRESSION

This is like creating a secret code. A variable is a symbol that stands for a number we don't know yet. An equation is a sentence with equal (=) signs that tells us something. A common mistake is reversing the order in subtraction or division. Here are some

> **EXAMPLE:** Tessa has 3 less marbles than Jesus, the equation could be $T = J - 3$.
>
> A meal costs $5 plus the price of the drink. We could write $M = 5 + D$.
>
> Joe worked half as hard as Phil. We would write $J = P/2$.

The same question could be stated in terms of Phil. Phil worked twice as hard as Joe $P = 2J$.

HOW FUNCTION TABLES WORK

A function table is like a machine: you put a number in, it does something to the number, and you get a number out. Imagine a function table that adds 2 to any number you put in. If you put in 3, you get out 5. The rule for that function would be Output=Input+2, for example $y = x + 2$. This would have to apply to all values in the table, so if $y = 3$ x would have to equal 1.

It's like a vending machine: you put in a dollar, and you get out a candy bar. If you put in three dollars you should know exactly how many candy bars will come out or the vending machine will not... function. Common mistakes include applying the wrong operation or the wrong order of operations.

HOW TO MATCH A FUNCTION TABLE TO A VERBAL EQUATION

This is like a detective game. You look at what the function table does to each input number, and you try to find the rule that explains it in words. The key is the same rule has to apply to all the values in the table in their given input and output pairs. If a function table adds 2 to any number, you could say "the output is the input plus 2". If you put in 2 and get out 4 the rule works. Then if you put in 10, but instead get out 20, you would know the rule is not +2, and then have to find a rule that worked for both pairs on input and output values. The rule in that case would be x2 as it works for both sets of values.

HOW TO SOLVE EXPRESSIONS WITH SUBSTITUTION

This is like trading snacks with your friend at lunch. If you know what a variable stands for, you can trade, or substitute the variable with its value. If $x = 5$, and you have the expression $x + 3$, you can swap the x with 5 to get $5 + 3 = 8$.

Let's try a more difficult problem, solve for x when **$3x+2y=10-3y$** when **$y=-2$**
Start by substituting in the value for y $3x+2(-2)=10-3(-2)$
Then simplify $3x-4=10+6$
Then combine the like terms $3x=10+6+4$
    $3x=20$
    $x=\dfrac{20}{3}$

WHAT ARE THE FOUR QUADRANTS OF THE X-Y CHART?

An X-Y chart, or a coordinate plane, is like a map. It has four parts (quadrants), and we can describe any point's location on this map using an ordered pair of numbers. The pair of numbers, or ordered pair, contain an x and y value similar to a function table. The x value always comes first and describes where the point falls on the horizontal axis. The y value is the second number and describes where the point will fall on the vertical axis. Both the x and y axis are layed on across each other forming perpendicular lines which creates 4 quadrants (I,II,III, IV.)

If a point is in the upper right corner, it's in the first quadrant. Here, both x (left or right) and y (up or down) are positive.

If a point is in the upper left corner, it's in the second quadrant. Here, the x (left or right) value is negative and the y (up or down) value is positive.

If a point is in the lower left corner, it's in the third quadrant. Here, the x (left or right) value is positive and y value (up or down) is negative.

If a point is in the lower right corner, it's in the fourth quadrant. Here, both x (left or right) and y (up or down) are negative.

It's like the game "Battleship" where you find your opponent's ships using grid coordinates. Common mistakes include reversing the x and y coordinates or confusing the signs of the coordinates as they contain negative and positive integers.

HOW TO GRAPH ORDERED PAIRS ON AN X/Y CHART

An ordered pair is like an address for a point on a map. The first number tells you how far to move right or left, and the second number tells you how far to move up or down. If the ordered pair is (3, 2), you move 3 spaces to the right and 2 spaces up from the center (0,0).

To continue the "treasure hunt" example, "take 3 steps forward and 2 steps to the left". This would be represented by the point (-2,3) as the x value (horizontal) always comes before the y value (vertical.) Common mistakes include reversing the x and y coordinates, or moving in the wrong direction.

WHAT GRAPHS REPRESENT A FUNCTION AND WHICH DO NOT. WHAT IS A FUNCTION?

A function is a special rule where each input has exactly one output. Graphs that pass the "vertical line test" (vertical line only hits the graph once) represent functions. If you can draw a vertical line anywhere on your graph and it only touches the graph once, it's a function! It's like assigning chores at home. Each person (input) is assigned a specific chore (output). No one has more than one chore!

WHAT IS THE DOMAIN AND RANGE OF A FUNCTION?

The Domain is all the x-values or input values of the function.

The Range of a function is all of the y-values or outputs of the function.

SOLVING AN EQUATION GIVEN AN ORDERED PAIR

An ordered pair is a "key" that unlocks an equation. When you know the values of x and y, you can put them into the equation to see if it's true. If you're given the ordered pair (2, 4) and the equation $y = 2x$, you can replace x with 2 and y with 4 to see that $4 = 2*2$, which is true. Common mistakes include substitution errors or miscalculations.

HOW TO FIND THE AREA OF A CIRCLE

The area of a circle is like the space it takes up on a piece of paper. We use a special number, pi (π), and the circle's radius (distance from the center to the edge) to find it. If a circle has a radius of 3, the area would be π times the radius squared (or multiplied by itself), which is $3*3*\pi = 9\pi$. Common mistakes include forgetting to square the radius or confusing diameter with radius. Reviewing the formula for the area of a circle ($\frac{\pi r}{2}$) and the difference between radius and diameter can be helpful.

HOW TO FIND THE AREA OF TRIANGLES

The area of a triangle is the space it covers up. If you have a triangular garden, the area would tell you how many square feet of soil you need to cover it. You can find it by multiplying the base (bottom edge) by the height (straight line from the base to the top point), and then dividing by 2. This is very similar to the formula for finding the area of a rectangle by multiplying the length times the width. The base and the height are similar. However, if you were to draw a diagonal line across a rectangle, it would split into two equivalent right triangles. The reason the formula for area of a triangle is $\frac{BH}{2}$ is because the triangle you are finding the area of is basically half a rectangle. If a triangle has a base of 4 and a height of 5, the area would be $\frac{(4*5)}{2} = 10$.

HOW TO FIND THE PERIMETER OF A RECTANGLE GIVEN ITS AREA AND THE LENGTH OF ONE SIDE

The perimeter of a rectangle is like a fence around a yard, it's the total distance around the outside. If we know the area (like how much space there is inside the yard) and the length of one side, we can find the other three sides. We already know that the opposite sides of rectangles are congruent, so we really only need to find the value of the adjacent side. This could be used to find the perimeter.

If a rectangle has an area of 20 and one side is 4, the other side must be $20 \div 4 = 5$. Then, to find the perimeter, we add up all the sides: $4 + 5 + 4 + 5 = 18$. This works because for rectangles A=LxW. So, if you know A and either L or W you can isolate the other variable. In our examples let's substitute A=20 and W=4. 20=(L)4. We then divide both sides by 4 to find the value of the unknown side, in this case the length. We would plug 4 and 5 into our perimeter of a rectangle formula P=2L+2W so P=2(5)+2(4) and we would find P=18. Common pitfalls might be miscalculations or confusion between perimeter and area. The key is knowing the perimeter and area formulas.

Area of a Rectangle: A=LxW

Perimeter of a Rectangle P=2L+2W

HOW TO FIND THE RULE OR EQUATION GIVEN A SET OF ORDERED PAIRS OR X/Y CHART

Imagine you're a detective trying to figure out a secret code. The x/y chart gives you clues (the ordered pairs), and your job is to figure out the secret code (the equation or rule). The trick is that the rule must work for every set of ordered pairs. Take the example (-2,0), (-1,2), (0,4), (1,6)

X	-2	-1	0	1
Y	0	2	4	6

Both the x/y chart and the set of ordered pairs are the same data. Look for patterns of what happens to the value input (x) and what comes out (y.) Finding the output value for y when x is 0 is often helpful. In this case we know that we must add 4 to x to get our output. Now let's take that information and apply it to the other ordered pairs. In the case of (1,6) the ordered pair's values are more than 4 apart, so they must be increasing some other way as well. Notice that our other x values are negative and result in output values that are less than the 4 we know we have to add. Also notice that the larger the negative value, the more the output decreases. It would seem we need to multiply our x value by something and then add 4. Let's try multiplying our x value by 2 and then adding 4. In the case of starting with 0, it doesn't matter as anything times 0 is still 0 that point will still work with our rule x2 + 4. How about substituting in x=1. (1)2+4=6 so that point works! How about x=-1? (-1)2+4=2 so that works as well. Our rule works for all of the points in the table so the equation of the line is y=2x+4.

HOW TO FIND THE Y-INTERCEPT OF A LINE

The y-intercept of a line is the point where the line crosses the y-axis. In the equation of a line y = 2x + 3, the y-intercept is the number by itself, so it's 3. Reviewing the formula y = mx + b (where b is the y-intercept) and the visual representation of y-intercept on a graph can be helpful. Try a few examples. What are the y-intercepts of the following lines?

$y = \frac{2}{3}x - 2$ y-intercept is -2   y=50x+25 y-intercept is 25

y=-7x y- intercept is 0 as there is no value, which is still on the graph at 0

You can also find the y-intercept by setting x to 0 and solving for y. See how to find the X-Intercept below.

HOW TO FIND THE X-INTERCEPT OF A LINE

The x-intercept is the spot where a line crosses the x-axis. A common mistake is confusing the x-intercept with the y-intercept or the slope.

If you have the equation of a line y = 2x - 3, you find the x-intercept by setting y to 0 and solving for x. (0)=2x-3

$$3 = 2x$$
$$\frac{3}{2} = X$$

So $x = \frac{3}{2}$, so the x-intercept would be $\frac{3}{2}$ or 1.5.

HOW TO FIND THE SLOPE OF THE LINE GIVEN TWO POINTS

The slope of a line is like a hill. It tells us how steep the line is. It's the vertical change (up or down) divided by the horizontal change (left or right).

We use the point slope formula y-y1=m(x-x1) which can be rewritten as m=$\frac{(y-y1)}{(x-x1)}$

If we have two points (2,4) and (3,6), the slope is the change in y (6 - 4 = 2) divided by the change in x (3 - 2 = 1), so the slope is 2. A common mistake is switching the order of the points or subtracting in the wrong order. Remember, the concept of rise over run!

WHAT IS UNDEFINED SLOPE?

An undefined slope is like trying to walk up a wall—it's so steep, it goes straight up! A line that goes straight up and down, like x = -12, has an undefined slope because it doesn't go left or right at all. Undefined slope is a term used for vertical lines in coordinate geometry. A common mistake is to think that the slope is 0, but 0 slope is for horizontal lines. This is quite the opposite, instead of being a perfectly flat path, it is instead running into a brick wall!

"18. Write the Equation of a Line Given Two Points in Slope-Intercept Form"

This is like a secret recipe to make a line. You need two ingredients: a slope (how steep your line is) and a y-intercept (where it starts on the y-axis).

If you have points (2,3) and (4,7), first find the slope using m=$\frac{(y-y1)}{(x-x1)}$

$$\frac{(7-3)}{(4-2)}$$

$$\frac{4}{2}$$

m=2

Then use one point and the slope to find the y-intercept,

We know y=mx+b, so we can change the equation to solve for b: b = y - mx

Now we substitute in the slope we solve for with one of our ordered pairs, say (2,3).

b = y - mx

b = (3) - m*(2) Now substitute in our slope value

b = (3) - (2)(2)

b = 3 - 4

b=-1

The equation is y = 2x - 1.

HOW TO FIND THE EQUATION OF A LINE THAT IS PARALLEL TO ANOTHER LINE

Imagine you are laying down a new track next to an existing railway line. You need to make sure the new track has the same slope so that the trains can run parallel to each other. Parallel lines are like train tracks—they never meet and they have the same steepness (slope). So $y = \frac{1}{3} = x - 3$ is parallel to $y = \frac{1}{3}x + 7$. You can find a line parallel to another line with just a single point as you already know the slope of both parallel lines.

If a line's equation is $y = 2x + 3$ and you want to find a line parallel to it that goes through point (1,4), since parallel lines have the same slope, the new line will also have a slope of 2. Plug the slope and point into the equation $y - y1 = m(x - x1)$

$$y - 4 = 2(x - 1)$$
$$y - 4 = 2x - 2$$
$$y = 2x + 2$$

HOW TO SOLVE FOR Y GIVEN X IN AN EQUATION

If given one of the values in an equation you may be able to solve for another variable, or at least simplify the equation. You can only solve for one variable at a time, just like in an experiment, you should only be changing the independent variable, and everything else should be constant. A common mistake might be incorrect calculation or not correctly substituting the given x value. Solve by plugging the given value into your equation with substitution.

If the equation is $y = 2x + 3$ and $x = 2$

$y = 2(2) + 3$

$y = 4 + 3$

$y = 7$

HOW TO WRITE AN ALGEBRAIC EQUATION GIVEN A WORD PROBLEM

This is like turning a story into a secret code. You take the words from a problem and turn them into numbers and letters that we can solve in math. If the problem says "There are four times as many dogs (D) as cats, and there are C cats," you can write the equation for the number of dogs as $D=4C$.

Imagine you have three less than half as many dollars as your brother because you spent it all on energy drinks. If you represent your money with Y, and your brother's money with B you could write the equation $Y=\frac{B}{2}-3$

WHAT ARE SUPPLEMENTARY ANGLES?

Supplementary angles are like best friends—they always add up to 180 degrees, which is like a straight line or a half-circle.

If one angle measures 110 degrees, its supplementary angle 70 because supplementary angles must add up to 180 degrees. So, 110 + X =180, so 180-110 = 70.

Imagine opening a book completely flat on a table. The angle it makes is 180 degrees. If one page is tilted a bit at an angle, say 95 degrees, the other page will make up the rest to still get 180 degrees. So 180-95= 85 degrees.

HOW TO MULTIPLY TWO OF THE SAME VARIABLE TOGETHER

If x = 3, then x*x = 3*3 = 9. In other words, x^2 = 9. This is easier to understand with constants like numbers, but in algebra symbols just represent numbers we don't know yet. However, they still have to follow some rules! In the above example x*x would not equal 2x as x+x=2x. Just like 2(3)=6 vs 3*3=9

This concept introduces you to the idea of exponents and powers. A common mistake is to think x*x = 2x, but it's actually x. Reviewing the meaning of squared and why x*x equals x^2 can help clarify this concept.

PRACTICE TESTS

PRACTICE TEST #1

Each Practice Test will have four sections. Each section will have 15 questions which you will have 10 minutes to answer.

Good Luck!

SECTION 1.1

1.) How far apart is -7 from 28 on a number line?

A) 14
B) 28
C) 35
D) -35

2.) Peter Piper is making pickles and has $\frac{2}{3}$ a cup of vinegar. If he needs $4\frac{1}{4}$ cups of vinegar for his pickle recipe, how many more cups does he need?

A) $3\frac{1}{2}$
B) $3\frac{7}{12}$
C) $3\frac{3}{4}$
D) $3\frac{11}{12}$

3.) $\dfrac{3\frac{3}{4}}{\frac{1}{4}}$

A) 12
B) 15
C) 3
D) $3\frac{1}{4}$

4.) $6 \times 10^2 \times 8 \times 10^4$

A) 48,000,000
B) 48,000,000,000
C) 480,000,000
D) 4,800,000

5.) $\dfrac{3}{18} - \dfrac{7}{90}$

A) $\frac{4}{90}$
B) $\frac{1}{10}$
C) $\frac{4}{45}$
D) $\frac{1}{15}$

6.) Yuvi buys a share of stock for $10. Unfortunately, it goes down to $7.50. What percent did it go down?

A) 75%
B) 25%
C) 10%
D) 7.5%

7.) Joe gets catering for his son's birthday party. The catering costs $500. He is busy cleaning the house, so has to pay an 8% delivery fee. How much does he have to pay for the catering and delivery fee together?

A) $508
B) $400
C) $580
D) $540

8.) $\dfrac{48}{72}$ what would you simplify by too put this fraction in simplest form

A) 24
B) 48
C) 12
D) 6

9.) $\dfrac{8}{12} - \dfrac{1}{3} + \dfrac{1}{6}$

A) $\dfrac{7}{12}$

B) $\dfrac{1}{2}$

C) $\dfrac{5}{12}$

D) $\dfrac{14}{12}$

10.) Home Depot sells bags of mulch for about $6 for 1.5 cubic feet. If you need to mulch 2 garden beds that need 30 cubic feet each of mulch, how much will it cost to buy all the mulch?

A) $240
B) $180
C) $360
D) $480

11.) $24 + 6 \times 3 + \dfrac{(-24)}{3}$

A) 22
B) 38
C) 10
D) 34

12.) Larry Page, one of the founders of Google, has a net worth of 81 Billion dollars. What is that number in scientific notation?

A) 8.1×10^9

B) 8.1×10^{10}

C) 8.1×10^{11}

D) 8.1×10^{12}

13.) Kira's cookie company is selling cookies at a swim meet. She finds the number of cookies she has sold doubles every 30 minutes. At Noon she had sold 6 cookies. How many cookies will she have sold by 2 PM?

A) 90 cookies
B) 96 cookies
C) 36 cookies
D) 72 cookies

14.) What are the next two terms in the sequence?

3, -12, 48, _____, _____

A) -60, 72
B) -96, 192
C) -192, 768
D) 192, 768

15.) At a baseball game, Hot Dogs cost 50% of a Hamburger. If you buy 2 Hot Dogs and 2 Hamburgers for $30, what does a Hot Dog cost?

A) $5
B) $4
C) $6
D) $10

SECTION 1.2

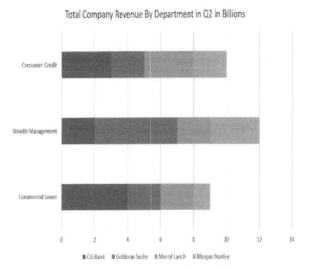

USE THE STACKED BAR CHART ABOVE FOR QUESTIONS 3-7.

1.) Ronan, Ben, Henry and Bert are trying to line up to go to recess. Ben can not stand next to Bert. Ronan can not be first in line today. Henry can not be last in line. Which shows the boys lined up from first to last so that they can go to recess?

A) Ben, Henry, Bert, Ronan
B) Ben, Bert, Henry, Ronan
C) Ronan, Bert, Henry, Ben
D) Bert, Ronan, Ben, Henry

2.) Chris is in a boat when he discovers there is a hole in it. He begins bailing out the boat while his brother starts rowing them back to shore. Every minute the boat takes on 50 cups of water. Chris can bail out the boat at a rate of 35 cups per minute. If it will take his brother 20 minutes to row the boat to shore, how much water will be in the boat?

A) 30 cups
B) 50 cups
C) 200 cups
D) 300 cups

3.) What is the mean amount of Commercial Loan Revenue of the four companies?
A) 2.25
B) 9
C) 2
D) 2.5

4.) What department brought in the most revenue across the four companies?

A) Consumer Credit
B) Wealth Management
C) Commercial Loans

5.) What department brought in the least revenue at Morgan Stanley?
A) Wealth Management
B) Commercial Loans
C) Consumer Credit

6.) What was the range of revenue from consumer credit?

A) 1
B) 3
C) 10
D) 2

7.) Which company made the most revenue on wealth management?

A) Merryl Lynch
B) Goldman Sachs
C) Morgan Stanley
D) Citi Bank

8.) USA, Mexico, Panama, and Jamaica are competing in the Concacaf Gold Cup in the group play stage. They must all play each other once. How many games will be played?

A) 4
B) 5
C) 6
D) 8

9.) 2 Tesla's and 2 BMW's cost $220,000. 2 Tesla's and 3 BMW's cost $280,000. How much does 1 Tesla cost?

A) $140,000
B) $70,000
C) $55,000
D) $50,000

10.) Listed below is a chart of how many chicken sandwiches Chick Fil A sold this week. The manager was not at work on Wednesday, but does know they sold 2,000 chicken sandwiches this week. How many chicken sandwiches were sold on Wednesday?

M	T	W	TH	F	S
300	240	???	360	350	600

A) 100
B) 150
C) 200
D) 250

11.) Students were asked to name their favorite continent in a survey:

Favorite Continent	Students
North America	15
South America	10
Europe	5
Asia	12
Africa	6
Australia	1
Antarctica	1

What percent of students chose Asia?
A) 12%
B) 18%
C) 20%
D) 24%

12.) At Marine Corp boot camp the trainees do 700 push ups in two days. If they continue at this pace, how many push ups will they do their first week of boot camp? Note that in the boot camp you do push ups every day of the week!

A) 1400
B) 2,450
C) 3,500
D) 4,900

13.) Students were asked to track how much time they spent doing the following activities outside of school.

Activity	% of Time Spent
Sleeping	70%
Reading	5%
Math HW	3%
TV/Video Games	15%
Play Outside	7%

If students have about 16 hours outside of school every day, about how many hours do they spend sleeping?

A) 24 Hours
B) 12 Hours
C) 11 Hours
D) 10 Hours

14.) Sherif has to carry five times as much weight in his backpack as his dad on a hike. If Sherif decides to put an equal amount of weight in both of their backpacks, both he and his dad will have 30 pounds in their backpacks. How many pounds does Sherif move into his dad's backpack?

A) 10 pounds
B) 15 pounds
C) 20 pounds
D) 25 pounds

15.) Bella studied for her IAAT for 3 hours on Monday. She studied twice as long on Tuesday as she did on Monday. On Wednesday she studied 4 hours less than she did on Tuesday. On Thursday she studied for half as long as she had on Wednesday. How many total hours did she study for her IAAT that week?

A) 9
B) 10
C) 12
D) 14

SECTION 1.3

1.) Write an equation for the relationship A is 5 more than half the value of B

A) $A = \dfrac{B}{5} + 2$

B) $A = 2B + 5$

C) $A = \dfrac{B}{2} + 5$

D) $A = 5B + 2$

2.) Danny's Dog Dojo charges $40 per session for its popular puppy training class. Danny has (S) Sessions this week. He has to pay a weekly rent of $250 for his business. Write an equation to show how many (D) Dollars Danny will make this week.

A) D=40S-250
B) D=250-40S
C) D=S-40+250
D) D=250S-40

3.) X + X + X = 24
X + X + Y = 26
What is the value of Y?

A) 10
B) 12
C) 14
D) 16

4.) Write two ordered pairs that would be in Quadrant IV.

A) (-2,2),(-3,3)
B) (-2,-2),(-3,-3)
C) (2,-2),(3,-3)
D) (2,2),(3,3)

5.) Which table shows the number of cookies (C) sold is 1 less than 3 times the number of Sodas (S) sold at the Swim Meet.

A)

C	1	2	3	4
S	2	5	8	11

B)

C	1	4	6	11
S	1	2	3	4

C)

C	2	5	7	4
S	1	2	3	4

D)

C	2	5	8	11
S	1	2	3	4

6.) Tracey has 4 fewer cookies than Casey. Stacey has twice as many cookies as Tracey. If Casey has (C) cookies, write an expression that shows how many cookies

A) S=2(C-4)
B) S=2T
C) S=C-4
D) $S = \dfrac{(C-4)}{2}$

7.) Which Scatter plot does not represent a function?

A)
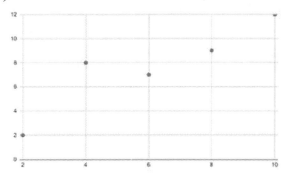

B)
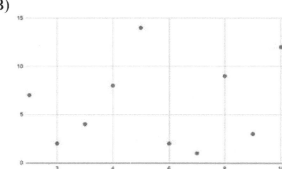

C)
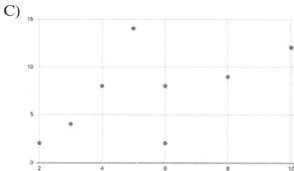

D)
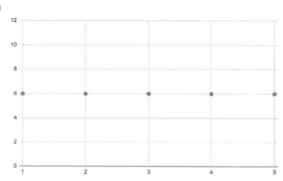

8.) Two premier league players average 12 goals a season. A third player is added to the average, and now the average goals scored per season is 18. How many goals did the third player score that season?

A) 30
B) 26
C) 18
D) 12

9.) The local Go Kart business charges a flat fee of $300 to rent their space for an hour plus $8 per person. Write an equation to show the total cost (C) for a group of (P) people to go Go Karting for an hour.

A) C + 300 = 8P
B) C = 300 + 8P
C) C = 8P(300)
D) C = 8P - 300

10.) Which line graph does not show a function?

A)
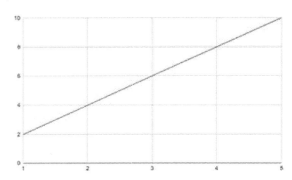

B)
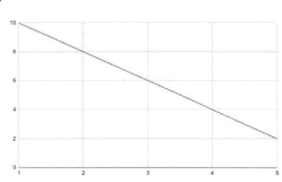

C)
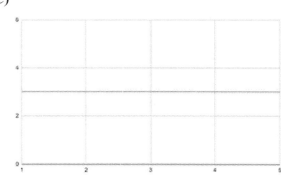

D)
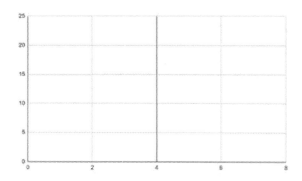

11.) Which ordered pair works for the equation y = 4x + 2

A) (4,2)
B) (-2,-6)
C) (2,6)
D) (-4,-12)

12.) Anaya's mom sends her with $15 to the local farmer's market to buy fruits. She can buy apples (X) for $1.25 and peaches (Y) for $2. Write an equation to show how much money she has after buying apples and peaches.

A) 15=1.25Y + 2X
B) 15=-1.25X - 2Y
C) 15=1.25X + 2Y
D) 15=1.25X - 2Y

13.) Copernicus has to be taken to the dog groomer, but when his owner takes him to the store, their credit card machine is down, and they are only accepting cash. His owner has $50 cash. The dog groomer charges $5 per dog plus $2.50/minute. How long can Copernicus be groomed for?

A) 20 minutes
B) 18 minutes
C) 16 minutes
D) 15 minutes

14.) If the radius of circle P is 4 inches and the radius of circle Q is 5 inches, given that Area of a Circle is πR^2, how much larger is circle Q than circle P?

A) 9π
B) 10π
C) 12π
D) 20π

15.) Birthday Parties at Chuck E Cheese cost (T) Total Dollars where X is the number of people attending the birthday and each person costs $18. You also have a coupon for $50 off. Write an equation for the cost of the party.

A) T=18 - 50X
B) T=18/X - 50
C) T=18X - 50
D) T=18X + 50

SECTION 1.4

1.) Simplify $3A + 2B - 3C$ If $A=2$ and $C=-2$
A) $2B +12$
B) $6+2B$
C) $2B-6$
D) $2B-6$

2.) If $Y=5$ and $Z=6$ what is the value of $4 + 3y - 4z$
A) -5
B) 5
C) 42
D) 43

3.) A Prime Energy drink has 200 mg which is 6 times as much caffeine as in a soda. Write an equation to show the amount of caffeine (C) in a soda.
A) $C= \frac{200}{6}$
B) $C= \frac{6}{200}$
C) $C= 200+6$
D) $C= 200-6$

4.) Solve for X: $8 > \frac{x}{8}$
A) $64<x$
B) $64>x$
C) $1<x$
D) $1>x$

5.) If x is an integer, which could not be a value for x^2
A) 1
B) 72
C) 81
D) 121

6.) Supplementary angles add up to 180 degrees. Angles O and P are supplementary. Angle P is 55 degrees. How many degrees is angle O?
A) 180
B) 155
C) 135
D) 125

7.) If $c^2 = a^2 + b^2$ and $a= 3$ and $b= 4$, what is the value of c?
A) 3
B) 4
C) 5
D) 6

8.) If $c^2= a^2 + b^2$ and $a= 6$ and $b= 8$, what is the value of c?
A) 9
B) 10
C) 12
D) 14

9.) The quotient of 42 and a number X is 7. Solve for X.
A) 5
B) 6
C) 7
D) 8

10.) Jake can do 5 sit ups in 8 seconds. Set up a proportion to show how many sit ups (S) he can do in a minute.
A) $60 = \frac{5s}{8}$
B) $\frac{5}{8} = \frac{s}{60}$
C) $\frac{5}{8}s = 60$
D) $\frac{5}{8} = s$

11.) What is the value of $x^4 + x$ if x=2?

A) 20
B) 18
C) 16
D) 10

13.) 3x + 12 = 48 Solve for x

A) 36
B) 24
C) 18
D) 12

15.) What is the domain of the given function (1,3), (2,4), (3,6), (4,8)

A) 1,2,3,4
B) 3,4,6,8
C) 1,3,2,4
D) 3,6,4,8

12.) Given $\frac{a}{2} = \frac{b}{3} = \frac{c}{6} = 3$
What is the value of a+b+c?

A) 36
B) 35
C) 34
D) 33

14.) Given x = 3 and y = -2.5 Solve -4y -2x

A) -4
B) -2
C) 2
D) 4

PRACTICE TEST #2

Each Practice Test will have four sections. Each section will have 15 questions which you will have 10 minutes to answer.

Good Luck!

SECTION 2.1

1.) Write $\dfrac{1}{8 \times 8 \times 8}$ in exponent form

A) 8^{-3}
B) 8^{3}
C) 24^{-3}
D) 512^{-3}

2.) Lucca is flipping coins 50 times. The ratio of heads to tails he flips is 1:1. How many heads does he flip?

A) 50
B) 100
C) $\dfrac{1}{50}$
D) 25

3.) $\dfrac{25}{0.5}$

A) 50
B) 5
C) 12.5
D) 125

4.) The latest Iphone is on sale for 25% off its retail price of $900. What is the sale price of the new Iphone?

A) 900
B) 225
C) 675
D) 775

5.) Eight new train stations in China are equally spaced in a straight line to each be the exact same distance apart. If the First station is 24 Km from the fourth station, how many Kms is it from the first station to the eighth and final station?

A) 42
B) 48
C) 56
D) 64

6.) There are 4 male chickens and 8 female chickens. Each female chicken produces 5 baby chickens. How many total chickens are there now?

A) 17
B) 52
C) 60
D) 28

7.) Theo's hero has 80 HP. Theo's hero has twice as much HP as Jack's hero. Ethan's hero has $\dfrac{1}{4}$ as much HP as Jack's hero. How many HP does Ethan's hero have?

A) 20 HP
B) 40 HP
C) 10 HP
D) 5 HP

8.) Evaluate 4^{-3}

A) $\dfrac{1}{64}$
B) $\dfrac{1}{12}$
C) -64
D) -12

9.) $3 + \dfrac{3}{3}$

A) $\dfrac{9}{3}$

B) 33

C) 4

D) $\dfrac{6}{3}$

10.) Gabrielle is harvesting cucumbers from the garden. Her dad pays her $0.50 per cucumber she picks. How many cucumbers will she need to pick to make $15?

A) 50

B) 30

C) 25

D) 15

11.) $\dfrac{5}{8} + \dfrac{1}{2} - \dfrac{3}{8}$

A) $\dfrac{9}{8}$

B) $1\dfrac{2}{8}$

C) $\dfrac{3}{4}$

D) $\dfrac{8}{8}$

12.) A bag full of StarBurst candy has 8 Red, 3 Yellow, 5 Orange and 4 Pink candies inside. If you really want pink, and stick your hand in the bag without looking, what is the probability you will pull out a pink StarBurst? Write your answer as a fraction, decimal and percent.

A) 20% $\dfrac{1}{5}$ or 0.2

B) 2% $\dfrac{1}{5}$ or 0.2

C) 25% $\dfrac{1}{4}$ or 0.25

D) 0.25% $\dfrac{1}{4}$ or 0.25

13.) Sana said she scored $\dfrac{47}{50}$ on her test. Anya said her test was 100 questions and she only missed 5. Henry said he scored 92%. Michael said he got 9 out of 10 correct. Who's percentage was highest?

A) Sana

B) Anya

C) Henry

D) Michael

14.) You eat either chicken or waffles for breakfast, lunch and dinner one day when you are on vacation. How many different possible combinations are there of what you could eat that day?

A) 2

B) 3

C) 5

D) 6

15.) $5(-6)(2)(-3)\left(\dfrac{1}{2}\right)$

A) 180

B) -360

C) -90

D) 90

SECTION 2.2

Use the double bar graph below of boys and girls favorite sports for questions 1-4.

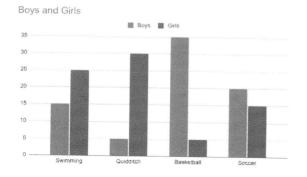

1.) About how many people in total were asked their favorite sport?

A) 130
B) 140
C) 150
D) 160

2.) How many more girls than boys preferred to play Quidditch?

A) 5
B) 10
C) 15
D) 25

3.) What percent of students surveyed preferred Girls Quidditch?

A) About 5%
B) About 10 %
C) About 20%
D) About 25%

4.) How many more students preferred swimming than soccer?

A) 5
B) 15
C) 20
D) 25

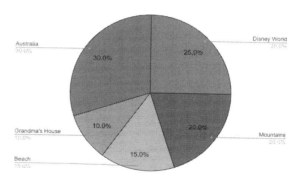

Use the Circle Graph above of students' favorite vacation destinations for questions 5-9.

5.) What percent of students surveyed picked Disney World?

A) 50%
B) 30%
C) 25%
D) 20%

6.) What two vacation spots added together are equivalent to Disney World?

A) Beach + Grandma's House
B) Mountains + Beach
C) Mountains + Grandma's House
D) Beach + Grandma's House

7.) According to the circle graph, what is the probability a student would select Grandma's House?

A) 0.15
B) 0.015
C) 0.01
D) 0.1

8.) If 40 people preferred vacationing in the Mountains, How many students were asked their favorite vacation spot?

A) 100
B) 150
C) 200
D) 250

9.) If 200 people were asked in the survey, how many more would have picked Australia or Disney World, then picked the Beach?

A) 80
B) 105
C) 125
D) 160

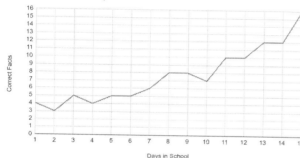

Use the line graph below of multiplication facts correct per day for questions 10-14.

10.) What is the mode number of multiplication facts answered correctly?

A) 12
B) 10
C) 8
D) All of the above

11.) What is the median number of multiplication facts answered correctly?

A) 6
B) 8
C) 7.5
D) 7

12.) How many more multiplication facts were answered correctly on day 15 than day 8?

A) 5
B) 6
C) 7
D) 8

13.) What is the range of the data set?

A) 12
B) 13
C) 14
D) 16

14.) Between what two days was there the greatest increase in multiplication facts answered correctly?

A) 14-15
B) 13-14
C) 9-10
D) 2-3

15.) Your teacher gets to school at 7:15 AM and begins working. She works until you go to lunch at 11 AM. At 11:30, after lunch she continues to work until 2:00 PM when you go to recess while she takes a 30 minute break. She then continues working when she picks you up from recess until she goes home exhausted, yet fulfilled at 5:30 PM. How many hours did your teacher work today?

A) 8 hours and 55 minutes
B) 9 hours and 15 minutes
C) 9 hours and 25 minutes
D) 9 hours and 55 minutes

SECTION 2.3

1.) Fill in the chart for the function y=6x-2

x	1	2	3	4
y				

A) 4, 10, 14, 20
B) 4, 10, 16, 22
C) 4, 8, 16, 20
D) 4, 10, 12, 24

2.) What is the next term in the sequence: 7, 9, 12, 16, _____

A) 18
B) 20
C) 21
D) 24

3.) What is the rule?

In	2	4	6	N
Out	-3	-1	1	????

A) 5 - N
B) N + 5
C) N - 5
D) $\frac{N}{5}$

4.) What is the X intercept of the line y = 3x + 9?

A) -3
B) 3
C) 9
D) -9

5.) Write an equation for one-fifth the number of band students (B) is twice the number of students in strings (S).

A) 2B = 5S
B) $\frac{B}{5} = 2S$
C) $\frac{B}{5} = \frac{S}{2}$
D) $\frac{B}{2} = 5S$

6.) $\frac{2}{3} < Y < \frac{3}{4}$

Give a possible value for Y

A) 0.5
B) 0.6
C) 0.7
D) 0.8

7.) Make a table that shows the number of Hot Dogs sold (H) is five less than four times the number of Cookies sold (C).

A)

C		2	3	4	5
H	3		7	11	15

B)

C		2	3	4	5
H	3		8	12	15

C)

C		2	3	4	5
H	4		9	12	16

D)

C		2	3	4	5
H	3		6	12	20

8.) What is the slope of a line 8x - 2y = 8
A) 8
B) -2
C) 4
D) -4

9.) If X is 12, what is Y?

x	1	3	5	8	12
y	4	12	20	32	

A) 36
B) 40
C) 48
D) 64

10.) When X=4, what is the value for y in the equation $y = \frac{3}{4}x - 2$

A) 1
B) 2
C) -2
D) 4

11.) Alfonso (A) scores four more goals than Mauricio (M). Write an equation for how many goals Mauricio scores.
A) M=A+4
B) M=$\frac{A}{4}$
C) M=4A
D) M=A-4

12.) Given the function, what is the value for B when A=8?

A	2	4	6	8
B	3	6	9	???

A) 10
B) 12
C) 16
D) 18

13.) Write the equation that matches the verbal statement: the number of car riders (C) is five more than six times the number of walkers (W).
A) C=W+5
B) C=6W+5
C) C=5W+6
D) C=6W-5

14.) If $y = \frac{1}{3}x - 2$ and x=15, what is the value of y?

A) 3
B) 4
C) 5
D) 6

15.) In their big win last night, the number of points scored by the Lakers (L) is ten less than double the number of points scored by the Magic (M.) If the Magic scored 62 points, how many points did the Lakers score?

A) 114
B) 116
C) 118
D) 120

Section 2.4

1.) If the absolute value of x is 5, give the two possible integers that x could be equal to

A) 25, -25
B) 5, 25
C) 5, -5
D) 5, $\frac{1}{5}$

2.) If the function A of B=B - A, what is 6 of 3?

A) 6
B) 3
C) -3
D) -6

3.) Lucas can drink $\frac{1}{4}$ of a chocolate milk in 1 minute. Write an expression that shows how many milks he can drink in X minutes.

A) 4X
B) $\frac{1}{4}$X
C) X-4
D) 4+X

4.) Write an equation to model 8 is decreased by a number, Q.

A) Q - 8
B) 8 - Q
C) 8Q
D) $\frac{8}{Q}$

5.) If x=5 what is the value of $\frac{x}{x^3}$

A) $\frac{1}{225}$
B) $\frac{1}{125}$
C) $\frac{1}{20}$
D) $\frac{1}{15}$

6.) If M=HS where M = total miles driven, H represents how many hours you drive, and S represents the speed you drive in the car. If the Amazon driver is driving for 8 hours at a constant speed of 65 miles per hour to reach your house, how many miles does he have to drive?

A) 520
B) 73
C) 57
D) 8.125

7.) If x = -8 then what is $\frac{x^2}{16}$

A) 4
B) -4
C) 8
D) -8

8.) If $\frac{x}{3}$ = 2 what is the value of 4x?

A) 8
B) 16
C) 24
D) 32

9.) Solve for G: -5G>20

A) G>-4
B) G<-4
C) G>-100
D) G<4

10.) What is 3B x 3B?

A) 9B
B) $3B^2$
C) $9B^2$
D) $6B^2$

11.) Rewrite the equation so that y is a function of x. $10y - 7 = 2x - 4y$

A) $y=\frac{1}{7}x+\frac{1}{2}$

B) $y=7x+\frac{1}{2}$

C) $y=7x+2$

D) $y=x+14$

12.) Simplify the expression: $(-12y)(-5y)$

A) $60y^2$

B) $60y$

C) $-60y$

D) $-60y^2$

13.) Solve the equation: $\frac{1}{5}(3-y)$ If $y=1.5$

A) 22.5

B) 2.25

C) 0.9

D) 0.3

14.) The sides of a quadrilateral are x, y, z and 5. If the perimeter of the quadrilateral is 22, what is the equation to solve for the perimeter of the quadrilateral in simplest form?

A) x+y+z=17

B) x+y+z=18

C) x+y+z=19

D) x+y+z=20

15.) Write the expression for the product of two times a number x and five times a number y.

A) (5X)(2Y)

B) 10XY

C) (2X)(5Y)

D) 2X+5Y

PRACTICE TEST #3

Each Practice Test will have four sections. Each section will have 15 questions which you will have 10 minutes to answer.

Good Luck!

Section 3.1

1.) Sammy is playing a game where his score doubles every 40 seconds. He started the game with 25 points he had scored previously. How many points will he have scored in 2 minutes?
A) 100 points
B) 150 points
C) 200 points
D) 400 points

2.) Today is Wednesday. What day will it be in 12 days?
A) Sunday
B) Monday
C) Tuesday
D) Wednesday

3.) 16 -(-8) - 8
A) 0
B) 16
C) 24
D) -16

4.) The perimeter of one of my raised bed gardens is 22 feet. If the bed is 7 feet wide, how long is the garden bed?
A) 15 Feet
B) 12 Feet
C) 4 Feet
D) 2 Feet

5.) What is the Least Common Multiple of 9 and 12?
A) 12
B) 9
C) 36
D) 3

6.) 7 x 3 + 10 x (25 ÷ 5)
A) 71
B) 155
C) 260
D) 371

7.) Phil practices his free throws every morning to make the high school basketball team. On Monday and Tuesday he averages 74%. He makes 65% on Wednesday. What is his free throw average those three days?
A) 71%
B) 70%
C) 69.5%
D) 46.3%

8.) Which is the largest integer?
A) $|-9|$
B) $\sqrt{24}$
C) -4^2
D) -10

9.) What is 0.002 x 0.41 in Scientific notation?
A) 8.2×10^{-5}
B) 8.2×10^{-4}
C) 8.2×10^{-3}
D) 8.2×10^{-2}

10.) The drywall I purchased at Home Depot to fix my basement was on sale for 15% off when I bought in bulk. The original price of the dry wall would have been $300. How much money did I save on the drywall?

A) $285
B) $72
C) $45
D) $15

11.) -8(-4)(-2)

A) 64
B) -64
C) -32
D) 32

12.) You loan your little brother some money, but you worry he won't pay you back. You charge him 800% interest on what you originally loaned him. Your parents find out that you claim he owes you $18. Since your math is obviously correct, how much did you originally loan your younger brother? Hint Amount owed= Amount Loaned + Interest you charge

A) $2
B) $4
C) $18
D) $20

13.) What is 7.2×10^4

A) 720,000
B) 72,000
C) 7,200
D) 720

14.) An inflatable bounce house has a length of 12 feet and a width that is 75% of the length. What is the area covered by the inflatable bounce house?

A) 900 feet square
B) 9 feet square
C) 144 feet square
D) 108 feet square

15.) $\frac{3}{4} \times \frac{3}{5} \times \frac{5}{9}$

A) $\frac{1}{2}$
B) $\frac{1}{4}$
C) $\frac{3}{4}$
D) $\frac{4}{9}$

SECTION 3.2

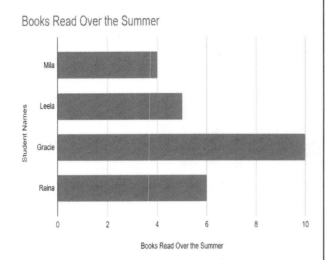

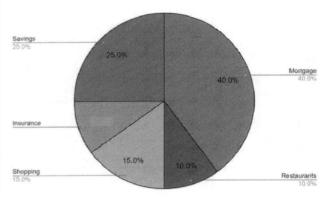

Use the bar graph above of Books Read Over the Summer for questions 1-3.

1.) How many books did Mila and Gracie read combined?

A) 14
B) 12
C) 10
D) 8

2.) How many books did Leela read?
A) 4
B) 4.5
C) 5
D) 6

3.) Assuming that summer was two months long, and Raina reads at the same pace all year, how many books will she read in a year?
A) 3
B) 6
C) 30
D) 36

Use the circle graph above of the Aguilars' monthly budget for questions 4-7. The Aguilar family makes about $12,500 per month.

4.) What % of the monthly budget goes toward insurance?

A) 5%
B) 10%
C) 15%
D) 20%

5.) How much is their mortgage payment every month?

A) $5,000
B) $4,750
C) $4,000
D) $50,000

6.) The Aguilar family is celebrating their daughter's Quinceanera later this year. They estimate it will take all of their savings for 2 months to throw an epic party! How much will they spend?

A) $5,500
B) $6,000
C) $6,250
D) $6,750

7.) The Aguilar's also want their daughter to start learning how to budget her money. They tell her that 25% of the money she earns working at the family business she should put into savings. Following her parent's advice, she put $75 into savings this month. How much money did she make this month?

A) $300
B) $275
C) $250
D) $225

Use the table below of Golden Corral Buffet and Grill Prices for questions 8-11.

Under 4	Under 12	Adults	Senior Citizens 65+
Free	10% off	$20	25% off

A group of friends and family going to Golden Corral is the following ages: 16, 17, 33, 31, 1, 2, 67, 80, 81, 46

8.) What is the mode of the ages of the group?
A) 10
B) 67
C) 81
D) No Mode

9.) What is the Median age of the group?
A) 1.5
B) 31
C) 32
D) 33

10.) How many family members pay full price at the buffet?
A) 10
B) 8
C) 5
D) 4

11.) What will the total bill be before taxes and tips?
A) $145
B) $130
C) $128
D) $120

Use the diagram below for questions 12-13.

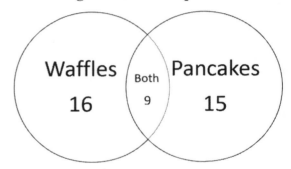

12.) How many total people were asked their favorite breakfast choice?
A) 9
B) 31
C) 40
D) 50

13.) What % of the people prefer ONLY Pancakes?
A) 40%
B) 37.5%
C) 30%
D) 15%

14.) Big Gus the squirrel eats 5 acorns every 20 seconds. How long will it take him to eat 14 acorns?
A) 7 Seconds
B) 30 Seconds
C) 56 Seconds
D) 70 Seconds

15.) What percent of students raised their hand to play Speedball?

15.) What percent of students raised their hand to play Speedball?

Game Time Choice	# of Students
Speedball	9
Night in the Museum	7
Follow the Leader	6
Heads Up 7 Up	3

A) 9%
B) 18%
C) 24%
D) 36%

SECTION 3.3

1.) Which is an equation for a line with undefined slope.

A) y=-2x
B) y=-2
C) x=2
D) y=$\frac{x}{2}$

2.) Omar, Bella, and Liam are arguing about who should get to be first in line. How many different combinations of line order can their teacher put them in?

A) 2
B) 4
C) 6
D) 8

3.) In your teacher's prize box, candy (C) is 2 for a ticket, and pencils (P) are 5 for a ticket. Write an algebraic expression to show how many tickets (T) it costs for 1 piece of candy and 1 pencil.

A) T= C+P
B) T=2C+5P
C) T= $\frac{C}{2}+\frac{P}{5}$
D) T= $\frac{C}{5}+\frac{P}{2}$

4.) What is the rule for the function?

a	b
1	0.5
2	1
4	2

A) B=$\frac{A}{2}$
B) B=2A
C) B=A+2
D) B=A-2

5.) Mr. Beast is a youtube influencer. He makes $100 a week in endorsements and $0.25 per like (L) on his videos. Write an equation that shows Mr. Beast's total Income (I) this week.

A) I=100L + 0.25
B) I=0.25L + 100
C) I=$\frac{L}{0.25}$ + 100
D) 100+0.25

6.) Listed below are the prices at the luxury car dealership's garage. The customer price is what they charge the customer to do the job. The labor price is what the dealership has to pay in wages to its employees to do the job. What car repair has the highest percent of labor cost relative to what the customer pays?

	Emissions	Oil Change	Flat Tire	Wind Shield Repair
Customer Price	50	100	150	400
Labor Price	10	30	30	100

A) Emissions Check
B) Oil Change
C) Flat Tire
D) WindShield Repair

7.) What is the slope of a line that passes through the points (-3,-3) and (0,3)?

A) 3
B) $\frac{1}{3}$
C) 2
D) $\frac{1}{2}$

8.) Solve y= 4(x^2) - 2x - 13 When x=5
A) 100
B) 77
C) 57
D) -3

9.) The ratio of boys who like french fries is four times as much as the girls. What is the ratio of girls to boys who like french fries?
A) 4:1
B) 1:4
C) $\frac{1}{4}$
D) $\frac{4}{1}$

10.) A rectangle has an area of 48ft^2. It's length is 12 feet. What is the Perimeter of the rectangle?
A) 32 ft
B) 42 ft
C) 48 ft
D) 64 ft

11.) Which does not represent a function?
A) (3,-3), (2,3), (4,5), (5,4)
B) (-2,-3), (-2,3), (3,5), (-3,4)
C) (3,-3), (2,-3), (4,-3), (5,-3)
D) (3,-3), (-2,3), (-3,5), (5,4)

12.) Pick the equation with the same slope as the line y=3x-3
A) y=$\frac{1}{3}$x-3
B) y=-3x-3
C) y=3x+9
D) y=$\frac{3}{x}$

13.) A square has an area of 64 inches. What is the square's perimeter?
A) 8 inches
B) 16 inches
C) 32 inches
D) 64 inches

14.) Casey leaves for her morning soccer game and drives 3 miles towards the game. She forgets she has to pick up the team shirts from Jordan, her injured teammate's house which is 5 miles the opposite direction of which she is driving. Once she gets the shirts, she drives 10 miles straight to the game. How far is the game actually from Casey's house?

A) 10 miles
B) 8 miles
C) 6 miles
D) 4 miles

15.) What is the equation of this function?

y	62	44	36
x	43	34	30

A) x= $\frac{y}{2}$ + 12
B) y= $\frac{x}{2}$ + 12
C) y= 2x - 12
D) x= $\frac{y}{2}$ - 12

SECTION 3.4

1.) Solve for Y: $\frac{1}{4}(y-2) = 10$

A) 48
B) 42
C) 4
D) 3

2.) Grayson's Athletic Shoe Company makes Profit (P)= Revenue (R) - Fixed Costs (C). Grayson's investors will give him a nice bonus if he can produce 1.5 Million in profit this month. If Grayson has fixed costs of 0.875 million this month, how much revenue does he need so his investors give him that sweet bonus?

A) 2.375
B) 0.625
C) 0.2375
D) 6.25

3.) -4y+9=-27

A) 9
B) -4.5
C) -9
D) 12

4.) If Angle Q is a right angle and the measure of Angle Q is 5X, what is the value of X?

A) 90
B) 36
C) 18
D) 5

5.) Giancarlo averages 3 goals every 4 games. Set up a proportion to show how many goals he scores in a 10 game season.

A) $\frac{3}{4} = \frac{x}{10}$
B) $\frac{4}{3} = \frac{x}{10}$
C) $\frac{4}{x} = \frac{10}{x}$
D) $\frac{3}{x} = \frac{10}{4}$

6.) Solve: -2z + 5 < 9

A) z<-2
B) z>-2
C) z>-7
D) z<-7

7.) The students get twice as rambunctious every day they have to stay inside for indoor recess. If they have had indoor recess four days in a row, how many times more rambunctious are the children in terms of R if their starting level of Rambunctiousness is set to R?

A) 4R
B) 8R
C) 16R
D) 64R

8.) How many factors does 24 have?

A) 2
B) 4
C) 6
D) 8

9.) If 3x=9 and x-y=15, what is the value of y?

A) -12
B) -6
C) 5
D) 6

10.) Five less than the product of three and a number (Z) is negative eight. Solve for Z

A) -5
B) -1
C) $\frac{3}{2}$
D) $\frac{13}{3}$

11.) The volume of a cone can be found using the formula $V = \frac{1}{3} H\pi r^2$

What is the volume of the cone if the radius is 4 inches and the height is 3 inches.

A) 12π
B) 16π
C) 24π
D) 48π

12.) $z^3 = 27$, what does $z^2 =$

A) 3
B) 6
C) 9
D) 18

13.) The sum of all the interior angles in all triangles is 180 degrees. In triangle ABC, Angle A measures 120, Angle B is 35, how many degrees is Angle C?

A) 50
B) 35
C) 25
D) 185

14.) If. $\frac{x}{3} + 2y = 11$ and $y = 2$, what is x equal to?

A) 55
B) 45
C) 35
D) 25

15.) You are on a long car trip with your family. Your younger brother constantly asks "are we there yet?" You explain to him that you had traveled x miles, 10 miles ago. You ask him how far you will have traveled in terms of x in 15 more miles. What is the answer?

A) x+25
B) x+15
C) x+5
D) x-5

PRACTICE TEST #4

Each Practice Test will have four sections. Each section will have 15 questions which you will have 10 minutes to answer.

Good Luck!

SECTION 4.1

1.) 85x12 + 12x15

A) 1200
B) 1020
C) 2100
D) 1002

2.) Seongwoo eats 47 chicken nuggets for breakfast, 36 nuggets for lunch, and 48 nuggets for dinner. He then promptly throws up 16 nuggets, because that is way too many chicken nuggets to eat in one day! How many nuggets will he actually digest?

A) 131
B) 115
C) 147
D) 127

3.) Tiberius can run a lap around the house in 12 seconds. How many laps can he run if he runs at a constant speed for 3 minutes?

A) 12 laps
B) 13 laps
C) 14 laps
D) 15 laps

4.) Potato Salad is normally $5.00 per pound. They made too much at Trader Joe's and put it on sale for 25% off. What is the sale price of the potato salad today?

A) $6.25 per pound
B) $2.75 per pound
C) $4.75 per pound
D) $3.75 per pound

5.) What is $\frac{3}{50}$ as a decimal and a percent?

A) 0.006 and 6%
B) 0.06 and 60%
C) 0.6 and 6%
D) 0.06 and 6%

6.) Mrs. Suarez's famous salsa calls for 3 tomatillos for every 7 tomatoes. If she bought 5 tomatillos at the store, how many tomatoes will she need?

A) 11 and $\frac{2}{3}$
B) 10 and $\frac{3}{7}$
C) 21
D) 35

7.) The number of Eagles fans at the Washington game was three times the number of Washington fans. There were 68,000 people at the game. How many Eagles fans showed up?

A) 17,000
B) 51,000
C) 34,000
D) 204,000

8.) Jolly Ranchers are sold in 12 packs for $3.50 at Walmart. To the nearest whole cent, how much are you paying per Jolly Rancher at Walmart?

A) 42
B) 4.2
C) 2.9
D) 0.29

9.) On your trip to Nebraska to go visit your friend who moved there, you bring four pairs of pants, three shirts, and two fancy hats. How many possible different outfits could you make, if you had absolutely no concern for how outlandish you might look?

A) 9
B) 12
C) 14
D) 24

10.) Safan runs 12 yards in 15 seconds at his soccer game. If he keeps running at this rate, how far will he run in 2 minutes?

A) 144
B) 96
C) 60
D) 29

11.) Deer can eat 12 plants in 15 seconds. If every plant has an average of 4 leaves, how many leaves will the deer eat in 1 minute?

A) 60 leaves
B) 31 leaves
C) 192 leaves
D) 720 leaves

12.) A class of 20 students want to buy 5 gallons of slime to dump on their unsuspecting teacher at field day. Each gallon of slime costs $16. If the cost of the slime is split evenly among all the students, how much will each student pay?

A) 4
B) 20
C) 80
D) 100

13.) $7^2 \times (\dfrac{7^4}{7^5})$

A) 7
B) 7
C) 7^2
D) 7^{11}

14.) The high temperature in Siberia yesterday was -20 degrees F. If every day for the next five days the temperature increases 4 degrees, what will the temperature be?

A) 0
B) -4
C) -20
D) 4

15.) Which inequality is false?

A) $9.01 \times 10^{-1} > 9.01 \times 10^{-2}$
B) $3.03 \times 10^{4} < 3.03 \times 10^{5}$
C) $1.11 \times 10^{8} < 1.11 \times 10^{-9}$
D) $1.23 \times 10^{-3} > 1.23 \times 10^{-4}$

SECTION 4.2

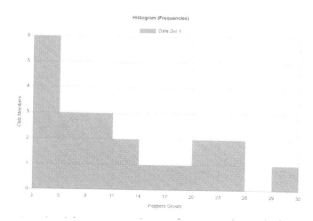

Use the histogram above for questions 1-4 about the Number of Peppers Grown by Garden Club Members.

1.) How many total students were in the gardening club?

A) 6
B) 18
C) 21
D) 32

2.) What is the range of peppers grown by gardening club members?

A) 32
B) 30
C) 6
D) 5

3.) If a garden club member is selected at random, what is the likelihood that person grew 8-11 peppers?

A) $\frac{1}{7}$
B) $\frac{8}{21}$
C) $\frac{1}{21}$
D) $\frac{8}{11}$

4.) $\frac{1}{3}$ of the garden club members grew so many peppers, they had to start giving their peppers away because they couldn't eat them all. How many peppers would be too many peppers?

A) 32
B) 21
C) 18
D) 14

Use the tabl;e below to answer questions 5-8. Emilia and Marta each asked 200 different people on the street if they prefer dogs or cats. They ask a total of 400 people. 160 people picked dogs, and 240 people picked cats. 180 of the people Marta talked to prefer cats.

	Dogs	Cats	Total
Emilia			200
Marta		180	200
Total	160	240	400

5.) How many people did Marta talk to prefer dogs?

A) 20
B) 40
C) 60
D) 80

6.) How many people did Emilia talk to that prefer dogs?

A) 140
B) 160
C) 180
D) 200

7.) How many people did Emilia talk to that prefer cats?

A) 40
B) 60
C) 80
D) 100

8.) What percent of the people preferred dogs?

A) 20%
B) 25%
C) 30%
D) 40%

Use the scatterplot below to answer questions 9-11.

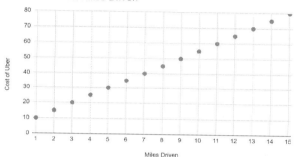

9.) An Uber ride of $60 will go how many miles?

A) 11
B) 12
C) 13
D) 14

10.) If you only have to go 6 miles, how much will the Uber cost?

A) 30
B) 30.5
C) 35
D) 40

11.) According to the scatter plot, as the length of the trip decreases, the cost of the trip ___Increases/Decreases___?

Use this information for Problems 12-14: The US Women's National Team scores the following number of goals in friendly matches this year: 3, 1, 2, 5, 10, 3

12.) What is the mean number of goals scored?

A) 3
B) 4
C) 5
D) 4.5

13.) What is the median number of goals scored?

A) 2
B) 3
C) 3.5
D) 5

14.) If the 10 goal game is removed from the data set, what is the new mean number of goals scored?

A) 3.5
B) 3
C) 2.8
D) 2.5

15.) $-5z+6+9z$

A) 20z
B) -20z
C) 14z+6
D) 4z+6

SECTION 4.3

1.) Pick the equation that matches the values in the X, Y chart

x	-6	-3	0	3
y	1	0	-1	-2

A) Y=3X-1
B) Y=$\frac{1}{3}$X+1
C) Y= -3X+1
D) Y= -($\frac{1}{3}$)X -1

2.) Bank of America is paying its first year accountants $90,000 per year. How much money would those accountants make per month?

A) $90,000
B) $9,000
C) $8,000
D) $7,500

3.) Hungry Howie's Hamburger food truck starts the day with $30 in cash. If they make $120 an hour in sales, write an equation that shows how much money (M) they have in (H) hours.

A) M= 120H +30
B) M= 120H(30)
C) M= $\frac{120}{H}$ +30
D) M= 30H +120

4.) Which set of points has the equation y= 2-x

A) (3,5), (4,7), (5,8), (1,3)
B) (-2,0), (-1,-1), (1,3), (2,4)
C) (1,1), (2,0), (3,-1), (4,-2)
D) (2,4), (4,6), (6,8), (8,10)

5.) Lebron hit 5 less three pointers than Steph last night. If Steph hit 13 three pointers, how many did Lebron hit?

A) 8
B) 6
C) 4
D) 2

6.)
X 1, 2, 3, 4
Y 3, 7, 11, ???
What goes in the ???

A) 12
B) 13
C) 14
D) 15

7.) The cost of uber eats to deliver to your house for dinner is $3.50 for the first mile and then $1.75 every mile after. If your desired restaurant is 8 miles away, write an equation that shows how much uber eats will cost you.

A 1.75 (7) + 3.50
B 3.50 (7) + 1.75
C 1.75 (8) + 3.50
D 3.50 (8) + 1.75

8.) To find the area of a triangle, multiply. $\frac{1}{2}$ BH. If the base of the triangle is 7, and the height is 4, what is the area of the triangle?

A) $21 ft^2$
B) $28 ft^2$
C) $7 ft^2$
D) $14 ft^2$

9.) Fill in the table to match the verbal equation the number of cats (x) is five less than four times the number of people (y).

x		3		11	
y		2	3	4	10

A) 3, 6, 11, 25
B) 3, 7, 11, 30
C) 3, 7, 11, 35
D) 3, 8, 11, 30

10.) What is the X-intercept of 3y+3x= 6

A) x=2
B) x=3
C) x=4
D) x=6

11.) Malik wants to have the coolest birthday ever, so he finds a Llama party rentals company, because who wouldn't want to have a llama party? It costs $250 for them to bring the llamas to your house and it costs $30 for every hour (H) they stay at your house. Write an equation showing the total cost (T) for this llama party.

A) T=30H +250
B) T=$\frac{30H}{250}$
C) T=30H(250)
D) T=30H -250

12.) Which of the following equations is parallel to the line y= -($\frac{1}{5}$)x +2

A) y=$\frac{1}{5}$x +2
B) y= - ($\frac{1}{5}$)x +10
C) y=5x +2
D) y=-5x -10

13.) At this week's soccer match 80% of the Strawberry Striker's passes were completed. Write an equation to show how many passes were completed (C) if they passed the ball 50 times.

A) C=0.8+50
B) C=$\frac{0.8}{50}$
C) C=(0.8)50
D) C=0.8-50

14.) The Friendly Local Game Store tracks how much money they make with the equation P=0.25S where P represents their total profits, and S represents the total sales of games. If the game store has $2,100 in sales on Saturday, how much money did they actually make in profit?

A) $600
B) $575
C) $525
D) $500

15.) When Eleanor, Leela and Alia go out to dinner at a fancy French restaurant, there is a fly in one of the girls' soup! The waiter is very embarrassed, and gives the girls 25% off of their whole bill (B). The girls plan to split the bill evenly. Which equation shows the correct amount (A) each girl should pay.

A) A=$\frac{B(1-0.25)}{3}$
B) A=$\frac{0.25(B)}{3}$
C) A=0.25(3B)
D) A=B($\frac{1}{3}$)0.25

SECTION 4.4

1.) Simplify: $\dfrac{3p^7}{9p^5}$

A) $3p^{12}$

B) $3p^2$

C) $3p^{-2}$

D) $\dfrac{p^2}{3}$

2.) Pick the equation that represents Six less than three times the difference of a number and five:

A) 6-3(x-5)

B) 6+3(5+x)

C) 3(x-5)-6

D) 3-6(x-5)

3.) If $5^x = 125$ what does x equal?

A) 3

B) 4

C) 5

D) 6

4.) $\dfrac{9x+3}{3}$

A) 4

B) 3x

C) 3x+1

D) 9x+1

5.) A is worth 3 times as much as B, and B is $\dfrac{1}{5}$ as valuable as C. C is worth $15. What is the value of A?

A) 5

B) 9

C) 15

D) 25

6.) The area of a circle is measured using Pi(r^2) If the radius of the circle is 0.5 inches, what is the area?

A) 0.225Pi

B) 0.25Pi

C) 0.5Pi

D) Pi

7.) $6x + \dfrac{1}{3}(3x+12)$

A) 6x+2

B) 7x+3

C) 7x+4

D) 9x+4

8.) Solve for x in the equation -4=14-2x

A) -18

B) -9

C) 9

D) 18

9.) If p=3 and q= -12, then what is the value of 3q +5p?

A) -21

B) 21

C) 51

D) -51

10.) The difference between a and b is 5. Which equation shows the algebraic statement correctly?

A) $\dfrac{a}{b}=5$

B) (B)(A)=5

C) A-B=5

D) A+B=5

11.) r-8>-12

A) r>-20
B) r>-4
C) r<-20
D) r<-4

12.) If 3R=6P=9, what is the value of R+P?

A) 1.5
B) 4.5
C) 9
D) 12

13.) If 5p - 5q = 50, what is the value of p-q?

A) p - q =50
B) p - q =40
C) p - q =10
D) p - q =5

14.) Simplify: 3 - 5z - 17 + 8z

A) 14 - 3z
B) 3z -14
C) 13z -14
D) 13z +20

15.) Claudia is trying to measure the height of the Eiffel Tower by measuring its shadow. Claudia knows she is 5 feet tall and measures her shadow at 20 feet. If she measures the Eiffel Tower's shadow as 2,000 feet, which proportional relationship would show the height of the Eiffel Tower?

A) $\frac{5}{20} = \frac{2000}{x}$

B) $\frac{20}{5} = \frac{2000}{x}$

C) $\frac{20}{2000} = \frac{x}{5}$

D) $\frac{5}{2000} = \frac{x}{2000}$

PRACTICE TEST #5

Each Practice Test will have four sections. Each Section will have 15 questions which you will have 10 minutes to answer.

Good Luck!

SECTION 5.1

1.) $\frac{5150}{50}$

A) 130
B) 103
C) 13
D) 0.13

2.) If you leave food inside your desk, the classroom might get ants. Fourteen people in the afternoon class get ants in their desks. It costs $50 to have exterminators clean out four desks. How much will it cost to have all the ants exterminated?

A) $700
B) $175
C) $150
D) $12.50

3.) Solve: $\frac{36}{\frac{9}{5}}$

A) 20
B) 65
C) 40
D) 4

4.) Jose shoots $\frac{8}{9}$ from the floor if you don't contest his shots. If he shoots 81 uncontested shots in practice, how many will he make?

A) 81
B) 72
C) 9
D) 8

5.) $\frac{24.48}{1.2}$

A) 2.4
B) 2.44
C) 20.4
D) 20.44

6.) Your grandpa falls asleep at 8:15 PM and wakes up at 6:55 AM to watch his favorite cable news show. How long was he asleep for?

A) 9 hours and 40 minutes
B) 9 hours and 50 minutes
C) 10 hours and 50 minutes
D) 10 hours and 40 minutes

7.) $9.1 + 0.0028 + 0.908$

A) 10.0108
B) 10.108
C) 100.1008
D) 10.0018

8.) What is the 7th number in this sequence? 2, 4, 8, 16, ???

A) 32
B) 64
C) 128
D) 256

9.) What is the Greatest Common Factor of 64 and 96?

A) 96
B) 64
C) 32
D) 16

10.) There are three blue gatorades and two yellow gatorade bottles. What is the probability of selecting a blue gatorade at random?

A) $\frac{1}{5}$
B) $\frac{2}{5}$
C) $\frac{3}{5}$
D) $\frac{4}{5}$

11.) The rectangular shaped floor of the classroom is 200% wider than it is long. If the length of the classroom is 20 feet, what is the perimeter of the classroom?

A) 60 Feet
B) 100 Feet
C) 120 Feet
D) 180 Feet

13.) Dave Chapelle holds a live comedy show and gives money to any audience member who's seat number has a 3 in it. There are 50 people at his show, sitting in seats numbered 1-50. How many people will he be paying?

A) 20
B) 14
C) 13
D) 12

15.) $\dfrac{56.7}{16.2}$

A) 0.85
B) 1.8
C) 3.5
D) 4.2

12.) 0.99×0.25

A) 0.2475
B) 247.5
C) 24.75
D) 2.475

14.) $(-5)^2 - 5(-3)$

A) -10
B) 25
C) 35
D) 40

SECTION 5.2

Use the line graph below for questions 1-4.

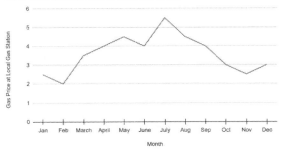

1.) How much did the price decrease from July to August?

A) $0.75
B) $2.00
C) $0.50
D) $1.00

2.) During which two months was there an increase in gas price?

A) June and July
B) August and September
C) September and October
D) October and November

3.) How much did the price increase from February to May?

A) $1.50
B) $2.00
C) $2.50
D) $3.00

4.) Between which two months was the greatest increase in price?

A) June to July
B) July to August
C) November to December
D) February to March

Shiv owns a food truck, and sells the following amounts of food during a given week. Use the chart for questions 5-8.

Day	Pizza	Donuts	Tacos	Fried Chicken
Mon	$125	$88	$76	$0
Tues	50	92	144	100
Wed	75	16	36	100
Thurs	0	44	144	100
Fri	150	200	100	100

5.) How much money did Shiv make selling Donuts that week?

A) $420
B) $430
C) $440
D) $450

6.) What food made Shiv the most money on Tuesday?

A) Pizza
B) Donuts
C) Tacos
D) Chicken

7.) What was the mean amount of Fried Chicken sold Monday through Friday?

A) $400
B) $100
C) $80
D) $75

8.) Shiv doesn't make his own Tacos, and buys them from his friend Jack and resells them. He spent $378 on the Tacos he sold this week, so how much profit did he make on his Tacos?

A) $112
B) $122
C) $134
D) $178

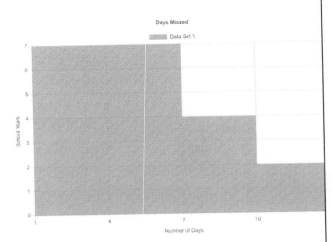

Use the histogram above to answer questions 9-12 about the Number of Snow Days Per School Year.

9.) What is the mode number of Snow Days Missed in a School Year?

A) 1-4
B) 4-7
C) All of the above
D) None of the above

10.) How many total years are included in the data set?

A) 13
B) 16
C) 18
D) 20

11.) What percent of school years did students have 7-10 snow days?

A) 4%
B) 15%
C) 20%
D 25%

12.) How many years did students miss 4-7 days of school?

A) 4
B) 5
C) 6
D) 7

13.) Jake catches crabs at the beach, and instead of eating them, attempts to race them! The Blue crab moves 100 inches across the beach. The Red crab moves 3 yards across the beach. The Orange crab then moves 8 feet and 2 inches. List the crabs in order from least to greatest of how far they moved across the beach. Note: 1 yard = 36 inches and 1 Foot = 12 inches.

A) Orange, Red, Blue
B) Orange, Blue, Red
C) Red, Orange, Blue
D) Red, Blue, Orange

14.) Five neighborhood children are picking cucumbers for Mrs. Suleiman while she is on a one week vacation and can not tend to her garden. She pays them $0.60 per cucumber they pick. The children tell Mrs. Suleiman she owes them $51. Assuming the children's math is correct, what is the mean number of cucumbers picked by each child

A) 17
B) 51
C) 85
D) 91

15.) 1, 2, 3, 4, 1, 2, 3, 4…
What is the 14th term in this sequence?

A) 1
B) 2
C) 3
D) 4

SECTION 5.3

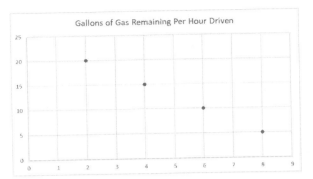

Use the above scatter plot to answer question 1.

1.) What type of relationship is shown in the scatterplot above?
A) Positive Relationship
B) Negative Relationship
C) No Relationship

2.) Write an equation for a line with the following points in slope intercept form.
(0,3), (-1,1)
A) y=2x+3
B) y=$\frac{x}{2}$ +3
C) y=2x-3
D) y=$\frac{x}{2}$ - 3

3.) Which set of ordered pairs represents a function?
A) (1,5), (3,7), (4,8), (3,3)
B) (-2,0), (-1,-1), (1,3), (-2,4)
C) (1,1), (2,0), (3,-1), (4,-2)
D) (2,4), (4,6), (8,8), (8,10)

4.) Which equation works for all of the x and y values in the table?

x		y
	1	3
	2	0
	3	-2
	4	-4

A) y=$\frac{1}{3}$x +2
B) y=2x +2
C) y=x+4
D) y=-2x+4

5.) A rectangle has a width of 4 feet and a length of 5 feet. If the width is increased by 25% and the length is increased by 20%, what is the new area of the rectangle?
A) 30 ft squared
B) 36 ft squared
C) 40 ft squared
D) 48 ft squared

6.) Which pair of points have a slope of $\frac{3}{2}$?
A) (7,2), (10,4)
B) (7,2), (9,5)
C) (3,2), (4,4)
D) (-3,-2), (0,0)

7.) For the points given, match the equation (-2,4), (0,1), (2,-2), (4,-5)
A) y= 2x +1
B) y= -2x +1
C) y= -($\frac{3}{2}$)x +1
D) y= -($\frac{2}{3}$)x +2

8.) If y=10, what is the value of x in the equation $y = \frac{3}{4}x - 2$

A) 20
B) 18
C) 16
D) 14

9.) Your parents are currently 4 times your little sister's age. If your sister is currently 10 years old, how old will your parents and your sister be when your sister is half their age?

A) 40
B) 50
C) 60
D) 80

10.) Write an equation for the mean or average of x, y and z is 15

A) $3(X+y+z) = 15$
B) $\frac{(X+y+z)}{3} = 15$
C) $15(X+y+z) = 3$
D) $(X+y+z) = 15$

11.) Which equation matches the table?

y	1	2	3
x	8	10	12

A) x=4x+4
B) y=8x
C) y=x+7
D) y=2x+6

12.) Which equation has an undefined slope?

A) y=2x
B) y=-3
C) x=-9
D) $y = \frac{1}{2}x$

13.) If the x intercept of y=mx+b is -1, what are the coordinates of the x intercept?

A) (-1,0)
B) (-1,-1)
C) (0,-1)
D) (1,0)

14.) DJ (J) has scored 7 less touchdowns than Derrick (R) this season. Match this to the equation that is equivalent to show how many touchdowns Derrick scored.

A) R=J-7
B) R=7-J
C) R+J=7
D) R=J+7

15.) Chewy.com, a pet supply website, offers better prices if you buy more pet food, up to a certain point. Chewy discounts their prices at a continuous unit rate, until you bulk order 800 containers of pet food. At that point they won't sell it to you any cheaper, and that is the best price you can get. Based on the chart below, what would Chewy's best price be for 800 containers of pet food?

Containers	25	50	100	200
Unit Price	$1.50	$1.35	$1.20	$1.05

A) $0.90
B) $0.75
C) $0.65
D) $0.50

SECTION 5.4

1.) Jackson can swim 50 yards in 1 minute. Which proportion shows how many yards (Y) Jackson swims in 90 seconds.

A) $\frac{1}{50} = \frac{90}{y}$

B) $\frac{60}{50} = \frac{y}{90}$

C) $\frac{50}{1} = \frac{90}{y}$

D) $\frac{60}{50} = \frac{90}{y}$

2.) $3(-2z+8)$

A) $6z+24$
B) $-6z+24$
C) $-2z+24$
D) $-2z-24$

3.) What is the value of z if $2y + 2z = 2x$ and $x = y +2$

A) 2
B) 4
C) 6
D) 8

4.) Which expression matches the following verbal statement: seven less than the quotient of six and five times a number, z.

A) $\frac{5}{6}z - 7$

B) $\frac{6}{5}z - 7$

C) $7 - \frac{6}{5}z$

D) $7 - \frac{5}{6}z$

5.) If A of B = $B^2 + \frac{1}{2}A$, what is the value of 4 of 5?

A) 45
B) 27
C) 18
D) 9

6.) Solve for z: $\frac{1}{3}(z+6)=5$

A) 21
B) 18
C) 11
D) 9

7.) Alex and Connor drop bread crumbs every 200 feet as they walk through the woods so they won't get lost. If they have to walk an unknown number of feet (U) to grandma's house, how many times will they drop bread crumbs, write an equation to show how many bread crumbs (B) they drop.

A) $B=200-U$

B) $U=\frac{B}{200}$

C) $B=200U$

D) $B=\frac{U}{200}$

8.) If $z-3 = -4$, What is $\frac{3-3z}{14}$

A) $\frac{3}{7}$

B) $\frac{1}{2}$

C) $\frac{3}{14}$

D) $\frac{8}{14}$

9.) Solve: $x^2\ 4 = 45$

A) 44 or -44
B) 11 or -11
C) 7 or -7
D) 5 or -5

10.) If A= 7 and B= -5, what is the value of $B^2 + 2AB$

A) 45
B) -45
C) 55
D) -95

11.) Solve the proportion: $\dfrac{8}{11} = \dfrac{Z}{99}$

A) z=792
B) z=9
C) z=27
D) z=72

12.) If A + B = 20 and A - B = 8, what are the values of A and B?

A) A= 20, B= 8
B) A= 16, B= 8
C) A= 12, B= 2
D) A= 14, B=6

13.) The volume of a cube is the length of any side s^3. If you need a cube shaped box to ship something you sold on eBay, and the box has a volume of 216 inches square, what is the longest length of any side of the object you can fit in the box?

A) 4 inches
B) 6 inches
C) 8 inches
D) 12 inches

14.) Solve for y, when $3y - 2z = -13$ if $z = 2$

A) -27
B) -20
C) -6
D) -3

15.) The perimeter of a rectangular backyard pool is 70 feet. If the pool is 15 feet longer than it is wide, what is the area of the minimum cover needed to cover the pool?

A) 500 feet squared
B) 250 feet squared
C) 170 feet squared
D) 85 feet squared

PRACTICE TEST #6

Each Practice Test will have four sections. Each section will have 15 questions which you will have 10 minutes to answer.

Good Luck!

SECTION 5.4

1.) How far apart is -12 from 2 on a number line ?
A) 10
B) 12
C) 14
D) -10

2.) Famous Chef Rachel Ray is baking cupcakes and has 2 and $\frac{3}{4}$ a cup of flour left. If she needs 5 and $\frac{1}{3}$ cups of flour for her cupcake recipe, how many more cups does she need?

A) $2\frac{1}{2}$
B) $2\frac{7}{12}$
C) $3\frac{7}{12}$
D) $8\frac{1}{12}$

3.) $\dfrac{4\frac{4}{5}}{\frac{2}{5}}$

A) 10
B) 12
C) 14
D) 16

4.) $8 \times 10^{-3} \times 9 \times 10^{2}$
A) 7.2×10^{-6}
B) 7.2×10^{-5}
C) 7.2×10^{-1}
D) 7.2×10^{1}

5.) $\frac{5}{6} - \frac{6}{72}$
A) $\frac{60}{72}$
B) $\frac{65}{72}$
C) $\frac{11}{12}$
D) $\frac{3}{4}$

6.) Swathi buys a share of stock for $20. Unfortunately, it goes down to $12.50. What percent did it go down?
A) 25%
B) 37.5%
C) 62.5%
D) 66%

7.) Jayden is renting out his AirBnB. His last guests made a huge mess, so he pays a cleaning company $800 to fix it up. He feels bad about how messy the job was so gives a 15% tip to the cleaners. How much was the tip?
A) $80
B) $100
C) $120
D) $920

8.) What is $\frac{56}{48}$ in simplest form?
A) $\frac{56}{48}$
B) $\frac{6}{7}$
C) $1\frac{1}{6}$
D) $\frac{4}{3}$

9.) $\dfrac{11}{12} - \dfrac{2}{3} + \dfrac{3}{4}$

A) 1
B) $\dfrac{3}{4}$
C) $\dfrac{1}{3}$
D) $\dfrac{1}{4}$

10.) Home Depot sells bags of concrete for about $7 for 2.5 cubic feet. If you want to expand your driveway, you calculate you will need 50 cubic feet each of concrete. How much will it cost to buy all the concrete?

A) $20
B) $100
C) $140
D) $350

11.) $-14 + 9(3) + \dfrac{(-28)}{7}$

A) 9
B) -9
C) -19
D) 19

12.) Many people have heard cats have 9 lives, but did you know they have an average of 70 million cells. What is that number in scientific notation?

A) 7×10^7
B) 7×10^8
C) 7×10^9
D) 7×10^{10}

13.) Dave is grilling pork chops at his family reunion BBQ. He finds the number of pork chops he has grilled doubles every 20 minutes. At Noon he had grilled 4 pork chops. How many cookies will he have grilled by 2 PM?

A) 32
B) 64
C) 128
D) 256

14.) What are the next two terms in the sequence?

108, 18, 3, _____, _____

A) $\dfrac{1}{2}, \dfrac{1}{12}$
B) $\dfrac{1}{6}, \dfrac{1}{12}$
C) $\dfrac{1}{2}, \dfrac{1}{6}$
D) $\dfrac{1}{2}, \dfrac{1}{18}$

15.) At a candy store, Sour Patch cost 50% of Fun Dip. If you buy 2 Sour Patch and 2 Fun Dips for $4.50, what does a Sour Patch cost?

A) $0.75
B) $1.25
C) $1.50
D) $3.00

SECTION 6.2

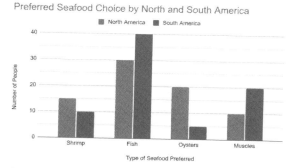

Use the double bar graph above to answer questions 1-4.

1.) How many people in total were asked their favorite seafood?

A) 125
B) 150
C) 160
D) 170

2.) How many more people from North America than South America preferred to eat Oysters?

A) 15
B) 10
C) 5
D) 20

3.) What percent of people preferred to eat Muscles?

A) 25%
B) 20%
C) 15%
D) 10%

4.) How many more people preferred Fish over Muscles

A) 55
B) 40
C) 15
D) 10

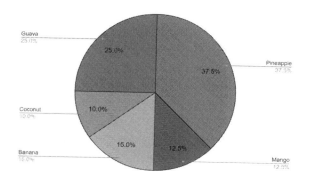

Use the pie chart above to answer questions 5-9.

5.) What percent of students surveyed picked Mango?

A) 12.5%
B) 25%
C) 50%
D) 62.5%

6.) What two favorite fruits added together are equivalent to Guava?

A) Mango and Banana
B) Banana and Coconut
C) Coconut and Pineapple
D) Coconut and Mango

7.) According to the pie chart, what is the probability a person would prefer Pineapple?

A) 25%
B) 33%
C) 37.5%
D) 45%

8.) How many people were asked their favorite tropical fruit if 25 people chose Mango?

A) 100
B) 150
C) 200
D) 225

9.) How many more people would you expect would prefer Banana or Guava, then prefer Coconut instead if you asked 1,000 people?
A) 150
B) 200
C) 250
D) 300

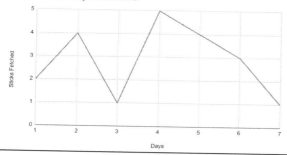

Use the line graph above to answer questions 10-14.

10.) What is the mode number of sticks fetched?
A) 1
B) 4
C) All of the above
D) None of the above

11.) What is the median number of sticks fetched?
A) 2
B) 3
C) 4
D) 5

12.) What is the mean number of sticks fetched?
A) 1.85
B) 2.50
C) 2.86
D) 3.50

13.) What is the range of the data set?
A) 1
B) 2
C) 3
D) 4

14.) Between what two days was there the greatest increase in the number of sticks fetched?
A) 3 and 4
B) 4 and 5
C) 5 and 6
D) 6 and 7

15.) Your mom gets to work at 8:15 AM and begins working. She works until she has a parent teacher conference with your teacher at 11 AM. At 11:30, after the conference she continues to work until 2:00 PM when she finally has a chance to eat lunch for a 45 minute break. She then continues working until she leaves work to take you to your various sporting events at 5:15 PM. How many hours did your teacher work today?
A) 9 hours 15 minutes
B) 9 hours
C) 8 hours 15 minutes
D) 7 hours 45 minutes

SECTION 6.3

1.) Which is an equation for a line with undefined slope.
A) y=2
B) x=2
C) y=2x
D) x=2y

2.) The Raiders, Chiefs, Chargers, and Broncos compete in the AFC West. How many different combinations can they finish their season when ordered first to last by winning records?
A) 8
B) 16
C) 24
D) 32

3.) In your teacher's prize box, a Gel Pen (P) is 5 tickets, and stickers (S) are 3 tickets. Write an algebraic expression to show how many tickets (T) it costs for 1 Gel Pen and 1 Sticker.
A) T=5P+3S
B) T=3P+5S
C) T=5P+5S
D) T=3P+3S

4.) What is the rule for the function?

a	b	
	2	0.8
	4	1.6
	5	2

A) B=0.8A
B) B=0.4A
C) B=4A
D) B=8A

5.) An S&P 500 Real Estate fund has increased $400 so far this year and pays a 5% dividend (D). Write an equation that shows the total return (T) for the index fund so far this year.
A) T=400(0.05D)
B) $T = \dfrac{D}{0.05 + 400}$
C) T=400D +0.05
D) T= 0.05D+400

6.) Listed below are the prices at a retail clothing store. The retail price is the price you pay at the store. The whole sale price is what the retailer has to spend to cover their costs for the item. What line of clothing has the highest percent of wholesale price relative to its retail price?

	H+M	Forever 21	Old Navy	The Gap
Retail Price	50	80	45	100
Whole-Sale Price	20	25	15	35

A) H+M
B) Forever 21
C) Old Navy
D) The Gap

7.) What is the slope of a line that passes through the points (2,1) and (0,3)?
A) $\frac{1}{2}$
B) -2
C) -
D) 3

8.) Solve $y=\frac{1}{2}x^2-3x-3$ When x=4
A) 11
B) -11
C) 7
D) -7

9.) The ratio of boys who like soccer is three times as much as the girls. What is the ratio of girls to boys who like soccer?
A) 1:3
B) 3:1
C) $\frac{1}{3}$:1
D) 1:$\frac{1}{3}$

10.) A rectangle has an area of 48ft^2. It's length is 6 feet. What is the Perimeter of the rectangle?
A) 14 feet
B) 20 Feet
C) 28 feet
D) 48 feet

11.) Which does NOT represent a function?
A) (3,4), (-4,-6), (6,8), (4,-10)
B) (3,4), (-4,5), (6,-6), (-3,-3)
C) (3,4), (-3,-6), (6,8), (-6,-10)
D) (3,4), (-3,-6), (9,8), (9,-10)

12.) Pick the equation with the same slope as the line y=3x-3
A) $y=\frac{1}{3}x-3$
B) y= 3x +27
C) y= x-3
D) $y=\frac{1}{3}x +27$

13.) A square has an area of 49 inches. What is the square's perimeter?
A) 7
B) 14
C) 21
D) 28

14.) On a camping trip you hike 3 miles towards the campsite. You then realize you made a wrong turn and hike 5 miles the opposite direction on the trail back to the trail marker. Once you get to the trail marker, you decide you would rather hike to the lake to go fishing and keep hiking 9 miles in the same direction you had been going. How far is the lake from the spot you realized you had made a wrong turn?
A) 17 miles
B) 14 miles
C) 8 miles
D) 7 miles

15.) What is the equation of this function?

y		6	10	14
x		1	3	5

A) $y=\frac{1}{5}x + 16$
B) y= 2x + 4
C) y= -x + 4
D) y= 4x - 4

SECTION 6.4

1.) Simplify: $\dfrac{3q^8}{9q^{10}}$

A) q^{-2}
B) $3q^2$
C) $\dfrac{1}{3}q^2$
D) $\dfrac{1}{3}q^{-2}$

2.) Pick the equation that represents seven less than four times the difference of a number and five:

A) -7(x-5)+4
B) 4(x-5)-7
C) 4(x)-12
D) -4(5-x)+7

3.) If $4^x = 256$ what does x equal?

A) 2
B) 3
C) 4
D) 5

4.) Simplify: $\dfrac{12x+4}{4}$

A) 4x + 1
B) 3x + 1
C) 8x
D) 3x

5.) A is worth 4 times as much as B, and B is $\dfrac{1}{4}$ as valuable as C. C is worth $15. What is the value of A?

A) $15
B) $60
C) $160
D) $240

6.) The area of a circle is measured using Pi(r^2) If the radius of the circle is 0.9 inches, what is the area?

A) 81Pi
B) 8.1Pi
C) 0.81Pi
D) 1.8Pi

7.) 2x $\dfrac{1}{6}$ (3x-18)

A) 2.5x-3
B) 1.5x-3
C) 20x-18
D) 1.5x+3

8.) Solve for y in the equation -13=14-3y

A) 9
B) -9
C) $\dfrac{1}{3}$
D) -$\dfrac{1}{3}$

9.) If p=10 and q= -12, what is the value of $\dfrac{1}{3}q + \dfrac{4}{5}p$?

A) 8
B) 4
C) -4
D) $\dfrac{4}{15}$

10.) The difference between a and b is 10. Which equation shows the algebraic statement correctly?

A) B=10-A
B) A=B-10
C) A=10-B
D) A-B=10

11.) -r+5>-10
A) r>-15
B) r>5
C) r<15
D) r<-5

12.) If 2R=4P=10, what is the value of R+P?
A) 2.5
B) 4.5
C) 7.5
D) 10

13.) If 8p - 8q = 56, what is the value of p-q?
A) 7
B) 8
C) 56
D) 64

14.) Simplify: 3 - 9z + 21 + 3z
A) 18 - 6z
B) 3z + 21
C) 6z + 18
D) 24 - 6z

15.) Aaliyah is trying to measure the height of a massive Redwood Tree by measuring its shadow. Aaliyah knows she is 5 feet tall and measures her shadow at 15 feet. If she measures the Redwood Tree's shadow as 1500 feet, which equation would show the height of the Redwood Tree?
A) $\frac{1}{15} = \frac{x}{1500}$
B) $\frac{x}{15} = \frac{1500}{5}$
C) $\frac{5}{15} = \frac{1500}{x}$
D) $\frac{5}{15} = \frac{x}{1500}$

PRACTICE TEST #7

Each Practice Test will have four sections. Each Section will have 15 questions which you will have 10 minutes to answer.

Good Luck!

SECTION 7.1

1.) Write $\frac{1}{12 \times 12 \times 12 \times 12}$ in exponent form

A) 12^4
B) $\frac{1}{-12^4}$
C) -12^4
D) 12^{-4}

2.) Cinderella picks petals off 86 flowers to determine if she will ever find true love. The ratio of flowers that result in true love, and those that end in being single forever is 1:1. How many of her flowers predict true love?

A) 86
B) 43
C) 50
D) 172

3.) $\frac{75}{0.5}$

A) 0.15
B) 15
C) 37.5
D) 150

4.) Takis are on sale for 25% off their retail price of $6.96 What is the sale price of the takis?

A) $8.70
B) 6.71
C) 5.22
D) 1.74

5.) The PE teacher puts out nine cones equally spaced in a straight line to each be the exact same distance apart. If the first cone is 18 feet from the fourth cone, how many feet is it from the first cone to the ninth and final cone?

A) 9 feet
B) 18 feet
C) 36 feet
D) 48 feet

6.) There are four 5th graders and seven 6th graders attending the 6th grade graduation party. Each 6th grader brings three additional friends because they are more social than 5th graders. How many total students will be at the graduation party?

A) 11
B) 14
C) 25
D) 33

7.) Ella scores 1000 points in Jeopardy. Ella's scores twice as many points as Sophia. Gracie scores $\frac{3}{5}$ many points as Sophia. How many points does Gracie score?

A) 1500
B) 500
C) 300
D) 100

8.) Evaluate 9^{-3}

A) $\frac{1}{729}$
B) $\frac{1}{27}$
C) -27
D) -729

9.) $-9 + \frac{9}{9}$

A) -9
B) -8
C) 0
D) 9

10.) Maya is cleaning dishes after dinner. Her dad pays her $0.25 per dish she washes. How many dishes does she need to wash to make $15?

A) 60
B) 30
C) 20
D) 375

11.) $\frac{7}{8} + \frac{3}{4} \quad \frac{5}{8} =$

A) 1 and $\frac{5}{8}$
B) 1
C) $\frac{5}{8}$
D) $\frac{3}{8}$

12.) A bag full of equally sized Dice has 9 Red Dice, 3 Green Dice, 6 Purple Dice and 2 Blue Dice inside. If you really want your lucky Blue Dice, and stick your hand in without looking, what is the probability you will pull out a Blue Dice? Write your answer as a fraction, decimal and percent.

A) 2%, 0.20, $\frac{2}{20}$
B) 10%, 0.10, $\frac{1}{10}$
C) 20%, 0.02, $\frac{2}{20}$
D) 20%, 0.2, $\frac{2}{20}$

13.) Mateo said he scored 48/50 on his test. Santiago said his test was 20 questions and he only missed 1! Beau said he scored 94%. Miles said he got 9 out of 10 correct. Who's percentage was highest?

A) Mateo
B) Santiago
C) Beau
D) Miles

14.) You eat either Tacos, Pizza, or Chicken Wings for breakfast, lunch and dinner one day when you are on vacation. How many different possible combinations are there of what you could eat that day?

A) 9
B) 6
C) 3
D) 1

15.) $-4(-6)(-2)(-3)(\frac{1}{6})$

A) -144
B) -24
C) 24
D) 144

SECTION 7.2

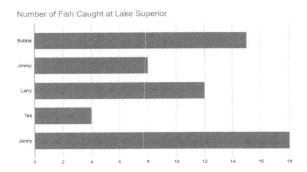

Use the bar graph above to answer questions 1-3 about the number of fish caught on day one of a fishing tournament on Lake Superior.

1.) How many fish did Jimmy and Larry catch combined?
A) 30
B) 26
C) 20
D) 4

2.) How many fish did Bubba catch?
A) 15
B) 14
C) 16
D) 14.5

3.) Assuming that the fishing contest goes for a whole week long, and Tex consistently catches fish at the same rate all week, how many fish will he catch?
A) 4
B) 18
C) 24
D) 28

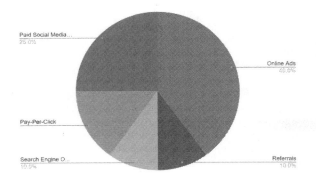

Use the Pie Chart above to solve questions 4-7 about Optimal Math's monthly marketing budget of $1,500 per month.

4.) What % of the marketing budget goes toward Pay-Per-Click (PPC)?
A) 20%
B) 15%
C) 10%
D) 5%

5.) How much do they spend on Online Ads every month?
A) $600
B) $750
C) $850
D) $900

6.) Optimal Math is looking to branch out into Math Board Games. They estimate it will cost the same as paying their Social Media Influencers for two years to afford the new venture! How much will they spend?
A) $10,000
B) $9,000
C) $4,500
D) $375

7.) Optimal Math is not convinced by their Search Engine Optimization (SEO) and compares it to their Online Adds. The Online ads produced $2,400 in sales. How much would SEO need to produce in sales at the same rate? (Note: We don't spend the same on SEO and Online Ads)
A) $150
B) $300
C) $600
D) $750

Reading a Table
Top Golf Admission Prices

Under 4	Under 12	Adults	Senior Citizens 65+
Free	10% off	$50	50% off

A group of friends and family are going to TopGolf for your dad's 50th Birthday. This is the group's ages: 12, 17, 53, 51, 1, 50, 67, 80, 81, 53

8.) What is the mode of the ages of the group?
A) 81
B) 50
C) 53
D) No Mode

9.) What is the Median age of the group?
A) 52
B) 53
C) 1
D) 50

10.) How many family members pay full price at TopGolf?
A) 3
B) 4
C) 5
D) 6

11.) What will the total bill be before taxes and tips?
A) $350.00
B) $375.00
C) $450.00
D) $475.00

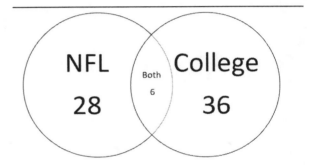

Use the Venn Diagram above of people's preferences for NFL vs College Football for questions 12-13.

12.) How many total people were asked their favorite type of Football?
A) 60
B) 64
C) 70
D) 76

13.) What % of the people prefer ONLY watching the NFL?
A) 40%
B) 34%
C) 28%
D) 4%

14.) Killer Whales can eat 12 fish every 30 seconds. How long will it take the whale to eat 60 fish?

A) 1.5 minutes
B) 2 minutes
C) 2.5 minutes
D) 3 minutes

15.) What percent of students raised their hand to play Speedball?

Game Time Choice	# of Students
Speedball	14
Night in the Museum	3
Follow the Leader	6
Pass the Beat	2

A) 14%
B) 28%
C) 50%
D) 56%

SECTION 7.3

1.) Pick the equation that matches the values in the X, Y chart

x	2	3	0	1
y	0	2	-4	-2

A) y=3x-2
B) y=-2x+4
C) y=2x-4
D) y=-2x+3

2.) Optimal Math pays its in person expert instructors $75 per hour. How much money would those instructors make working 40 hours per week?
A) $3,000
B) $1500
C) $1150
D) $115

3.) Tot Boss food truck, one of the top rated food trucks in America starts the day with $150 in cash. If they make $440 an hour in sales, write an equation that shows how much money (M) they have in (H) hours.
A) M=440H+150
B) M=150H-440
C) M=440H-150
D) M=150H+150

4.) Which set of points has the equation y= 5-x
A) (0,5), (4,7), (5,0), (2,3)
B) (5,0), (-1,-1), (0,5), (2,3)
C) (1,1), (5,0), (5,0), (2,3)
D) (1,4), (0,5), (5,0), (2,3)

5.) Luka hit 6 less 3 pointers than Damian last night. If Damian hit 15 3-pointers, how many did Luka hit?
A) 21
B) 18
C) 11
D) 9

6.)
X 3.....6....9....12
Y 15..30...45…???
What goes in the ???
A) 50
B) 55
C) 60
D) 65

7.) The cost of uber eats to deliver to your house for dinner is a flat $3.50 fee and then $2.75 for every mile during busy times. If your desired restaurant is x miles away, write an equation that shows how much uber eats will cost you.

A) C= 3.5x + 2.75
B) C=2.75x + 3.5
C) C=2.75(3.5)x
D) $C=\frac{2.75}{3.5x}$

8.) To find the area of a triangle, multiply $\frac{1}{2}$ 12BH. If the base of the triangle is 8, and the height is 5, what is the area of the triangle?
A) 40
B) 20
C) 24
D) 13

9.) Fill in the table to match the verbal equation the number of hats (h) is four less than three times the number of bald men (b).

h		0	2	
b	-4		-1	2	8

A) h=0, -2, 2, 6
B) h=0, 1, 2, 4
C) h=0, -1, 2, 4
D) h=0, 1, 2, 6

10.) The number of three point shots made (T) is typically four less than the number of free throws made (F). Which table matches this verbal equation?

A) F=T-4
B) T=F-4
C) T-F=4
D) T=4F

11.) Optimal Math is looking to host a meet and greet for new clients at a nice hotel. The venue deposit is $750 and $250 for every hour (H) the event lasts. Write an equation showing the total cost (T) for this meet and greet event.

A) T=750+250+H
B) 750T=250H
C) T=250H-750
D) T=750 +250H

12.) Which of the following equations is parallel to the line $y= -\frac{1}{5} x +2$

A) $y=-\frac{1}{5}x-7$
B) y=-5x+2
C) $y=\frac{1}{5}x+2$
D) y=5x-2

13.) At this week's soccer match 75% of the Strawberry Striker's passes were complete. Write an equation to show how many passes were completed (C) if they passed the ball a total of 50 times.

A) 50=0.75C
B) C=0.75(50)
C) 50=C-0.25
D) C=0.75+50

14.) The Friendly Local Game Store tracks how much money they make with the equation P=0.25S where P represents their total profits, and S represents the total sales of games. If the game store has $3,500 in sales on Saturday, how much money did they actually make in profit?

A) $3500
B) $2625
C) $1000
D) $875

15.) When Eleanor, Leela and Alia go out to get their nails done for Leela's birthday. Since it is her birthday, they are given a 33% discount of the total bill (B). The girls plan to split the bill evenly. Which equation shows the correct amount (A) each girl should pay.

A) A=0.67B+3
B) A=0.33B+3
C) $A=\frac{0.33B}{3}$
D) $A=\frac{0.67B}{3}$

SECTION 7.4

1.) Baby Cillian can swim 25 yards in 1 minute. Which proportion shows how many yards (Y) Cillian swims in 90 seconds.

A) $\frac{1}{25}=\frac{90}{y}$

B) $\frac{1}{25}=\frac{y}{1.5}$

C) $\frac{1}{25}=\frac{1.5}{y}$

D) $\frac{25}{1}=\frac{1.5}{y}$

2.) $4(-4z+\frac{1}{2})$

A) 16z+2
B) -16z+2
C) -16z-2
D) 16z-2

3.) What is the value of z if 3y + 2z = 3x when x =1 and y= -1

A) $\frac{2}{3}$
B) $-\frac{2}{3}$
C) 2
D) 3

4.) Which expression matches the following verbal statement: five less than the quotient of six and seven times a number, u.

A) $\frac{6}{7}u-5$
B) $\frac{5}{6}u-7$
C) $\frac{7}{5}u-6$
D) $5-\frac{6}{7}u$

5.) If B of A = $B^2+\frac{1}{2}$ A, what is the value of 3of4?

A) 7
B) 11
C) 12
D) 13

6.) Solve for z: $\frac{2}{3}(z+9)=5$

A) 1
B) -1
C) $-\frac{2}{3}$
D) $-\frac{3}{2}$

7.) Finn Robertson is enjoying a nice day out on the lake. He knows his boat holds 25 gallons of gas, but he also knows he used up an unknown number (U) of gallons already today. If his boat uses 2 gallons of gas per hour, how many hours (H) does he have before his boat runs out of gas?

A) 25=2H+U
B) U=2H+25
C) 25=2U+H
D) 25=2H-U

8.) If z+1 = -4, What is $5-\frac{5-2z}{15}$

A) $-\frac{1}{5}$
B) $-\frac{1}{3}$
C) 1
D) -1

9.) Solve: $x^2-4=77$

A) 9
B) 8
C) 7
D) 6

10.) If A= 3 and B= -5, what is the value of $B^3 + \frac{1}{3}AB$

A) 225
B) -65
C) -125
D) -130

11.) Solve the proportion: $\frac{9}{15} = \frac{r}{20}$

A) r=64
B) r=56
C) r=16
D) r=12

12.) If A + B = 24 and A - B = 6, what are the values of A and B?

A) A= 12, B= 12
B) A= 18, B=6
C) A= 24, B= 6
D) A= 15, B=9

13.) The volume of a cube is the length of any side s^3. If you need a cube shaped box to ship something you sold on ebay, and the box has a volume of 343 inches square, what is the longest length of any side of the object you can fit in the box?

A) 7 inches
B) 8 inches
C) 9 inches
D) 10 inches

14.) Solve for y, when 2y - 2z = 2 when z=-4

A) -6
B) -3
C) 3
D) 6

15.) The perimeter of a rectangular backyard pool is 100 feet. If the pool is 20 feet longer than it is wide, what is the area of the minimum cover needed to cover the pool?

A) 400 feet squared
B) 525 feet squared
C) 600 feet squared
D) 625 feet squared

PRACTICE TEST #8

Each Practice Test will have four sections. Each Section will have 15 questions which you will have 10 minutes to answer.

Good Luck!

SECTION 8.1

1.) Jin is playing a game where his score doubles every 20 seconds. He started the game with 5 points he had scored previously. How many points will he have scored in 2 minutes?

A) 160
B) 320
C) 640
D) 1600

2.) Today is Friday. What day will it be in 11 days?

A) Tuesday
B) Wednesday
C) Thursday
D) Friday

3.) -6 -(-19) - 8

A) -33
B) -21
C) -14
D) 5

4.) The perimeter of one of my raised bed gardens is 26 feet. If the bed is 7 feet wide, how long is the garden bed?

A) 5 Feet
B) 6 Feet
C) 7 Feet
D) 8 Feet

5.) What is the Least Common Multiple of 8 and 12?

A) 4
B) 16
C) 24
D) 48

6.) 8 x 3 + 10 x (36 ÷ 6)

A) 84
B) 96
C) 144
D) 204

7.) Arjun practices his free throws every morning to make the high school basketball team. On Monday and Tuesday he averages 84%. He makes 85% on Wednesday. What is his free throw average those three days

A) 84
B) $84 \frac{1}{2}$ %
C) $84 \frac{1}{3}$ %
D) 83

8.) Which integer has the largest absolute value?

A) -9
B) 0
C) 7
D) 8

9.) What is 0.00002 x 0.41 in Scientific notation?

A) 8.2×10^{-6}
B) 8.2×10^{-7}
C) 8.2×10^{-8}
D) 8.2×10^{-9}

10.) The drywall I purchased at Home Depot to fix my basement was on sale for 15% off when I bought in bulk. The original price of the dry wall would have been $250. How much money did I save on the drywall?

A) $25.75
B) $37.50
C) $38.25
D) $40.50

11.) $-8(-\frac{1}{4})(-12)$

A) -384
B) -144
C) 24
D) -24

12.) You loan your little brother some money, but you worry he won't pay you back. You charge him 500% interest on what you originally loaned him. Your parents find out that you claim he owes you $24. Since your math is obviously correct, how much did you originally loan your younger brother? Hint Amount owed= Amount Loaned + Interest you charge

A) $100
B) $29
C) $4
D) $2

13.) What is -8.6×10^{-5}

A) -860,000
B) -8,600,000
C) $\frac{1}{860,000}$
D) $\frac{1}{8,600,000}$

14.) An inflatable bounce house has a length of 20 feet and a width that is 80% of the length. What is the area covered by the inflatable bounce house?

A) 16 feet square
B) 36 feet square
C) 320 feet square
D) 400 feet square

15.) $\frac{3}{2} \times \frac{4}{5} \times \frac{15}{9}$

A) $\frac{1}{2}$
B) 2
C) $3\frac{1}{2}$
D) $\frac{16}{3}$

SECTION 8.2

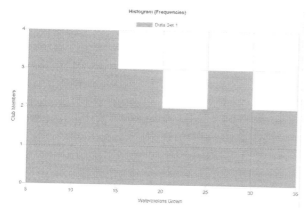

Use the histogram above about Number of Watermelons Grown by Garden Club Members for questions 1-4.

1.) How many total students were in the gardening club?

A) 4
B) 35
C) 16
D) 18

2.) What is the range of melons grown by gardening club members?

A) 30
B) 25
C) 4
D) 3

3.) If a garden club member is selected at random, what is the likelihood that person grew 10-15 melons?

A) 11%
B) 22%
C) 33%
D) 44%

4.) $\frac{1}{9}$ of the garden club members grew so many melons, they had to start giving their melons away because they couldn't eat them all. How many melons would be too many melons?

A) 2
B) 3
C) 4
D) 5

Reading a Table

Haleigh and Ava each asked 200 different people on the street if they prefer pie or cake. They ask a total of 400 people. 160 people picked cake, and 240 people picked pie. 180 of the people Ava talked to prefer pie.

	Cake	Pie	Total
Haleigh			200
Ava		160	200
Total	180	220	400

5.) How many people did Ava talk to prefer cake?

A) 20
B) 40
C) 60
D) 80

6.) How many people did Haleigh talk to that prefer cake?

A) 100
B) 120
C) 140
D) 160

7.) How many people did Haleigh talk to that prefer pie?

A) 20
B) 40
C) 60
D) 80

8.) What percent of the people preferred pie?

A) 50%
B) 55%
C) 60%
D) 65%

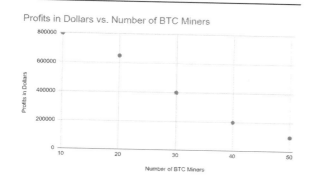

For questions 9-11 use the scatter plot above of Number of BTC Miners Compared to their Mining Profits

9.) How many miners are there making $200,000?

A) 40
B) 30
C) 50
D) 45

10.) If there are about 45 miners competing, how much profit would you expect to make (Put in between intervals or change problem to fit intended purpose)

A) $100,000
B) $150,000
C) $200,000
D) $250,000

11.) According to the scatter plot, as the number of miners decreases, the BTC mining profits (Increases/Decreases) ?

A) Increases
B) Decreases

12.) The local parking police issue the following number of tickets a day: 23, 21, 22, 25, 10, 23
What is the mean number of tickets issued?

A) 124
B) $21\frac{3}{4}$
C) $20\frac{2}{3}$
D) $20\frac{1}{2}$

13.) What is the median number of tickets issued?

A) 22
B) 22.5
C) 23
D) 23.5

14.) If the 10 ticket day is removed from the data set, what is the new mean number of tickets issued?

A) 22.2
B) 22.4
C) 22.6
D) 22.8

15.) -8z+8+4z

A) -4z+8
B) 4z+8
C) 4z-8
D) -4z-8

SECTION 8.3

1.) What type of relationship is shown in the scatterplot below?

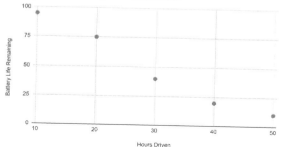

A) Positive Relationship
B) Negative Relationship
C) No Relationship

2.) Write an equation for a line with the following points in slope intercept form. (0,-1), (3,0)

A) $y=\frac{1}{3}x-1$
B) $y=\frac{2}{3}x-1$
C) $y=3x-1$
D) $y=2x-1$

3.) Which set of ordered pairs represents a function?

A) (0,1), (1,0), (-1,2), (2,-2)
B) (-2,0), (-2,-1), (-2,3), (-2,4)
C) (1,1), (2,0), (3,-1), (1,-2)
D) (2,4), (4,6), (8,8), (8,10)

4.) Which equation works for all of the x and y values in the table?

x	y
0	2
3	3
6	4
9	5

A) $y=-\frac{2}{3}x+2$
B) $y=-\frac{1}{3}x+2$
C) $y=\frac{2}{3}x+2$
D) $y=\frac{1}{3}x+2$

5.) A rectangle has a width of 8 feet and a length of 10 feet. If the width is increased by 75% and the length is increased by 20%, what is the new area of the rectangle?

A) 168 ft squared
B) 144 ft squared
C) 96 ft squared
D) 80 ft squared

6.) Which pair of points have a slope of $\frac{3}{1}$?

A) (1,1), (2,2)
B) (7,4), (9,5)
C) (0,2), (0,3)
D) (1,5), (0,2)

7.) For the points given, match the equation (-1,-5), (0,-3), (2,1), (5,7)

A) $y=\frac{1}{2}x-3$
B) $y=2x-3$
C) $y=x-3$
D) $y=\frac{2}{3}x-3$

8.) If y= -5, what is the value of x in the equation $y=\frac{3}{4}X-2$

A) 4
B) 3
C) -3
D) -4

9.) Your parents are currently 5 times your little sister's age. If your sister is currently 7 years old, how old will your parents be when your sister is half their age?

A) 60
B) 58
C) 56
D) 54

10.) Write an equation for the average of x, y and z is 7

A) $x+y+\frac{z}{3}=7$
B) $\frac{3xyz}{7}$
C) $\frac{3x+y+z}{7}$
D) $\frac{x+y+z}{7}=3$

11.) Which equation matches the table?

y	-3	3	1
x	0	-3	-5

A) y= x-1
B) y= x-3
C) $y=\frac{1}{2}x-3$
D) $y=-\frac{1}{2}x-2$

12.) Which equation has a slope of zero?
A) y=-1
B) y=-x
C) x=-1
D) y=x

13.) If the x intercept of y=mx+b is -17, what are the coordinates of the x intercept?
A (-17, 0)
B (0,-17)
C (17,0)
D (0,17)

14.) DJ (J) has scored 2 less goals than Emily (E) this season. Match this to the equation that is equivalent to show how many goals Emily scored.

A) J=E+2
B) E-2=J
C) E-J=2
D) E=2-J

15.) Optimal Math offers discounts if you buy their workbooks in bulk for your friends who also love Math :) They discount their prices at a continuous unit rate, until you bulk order 10,000 books. At that point we won't sell it to you any cheaper, and that is the best price you can get! Based on the chart below, what would our best price for 10,000 workbooks be?

Books	1	10	100	1000
Unit Price	$29.99	$26.49	$24.99	$22.49

A) $21.49
B) $20.99
C) $20.49
D) $19.99

193

SECTION 8.4

1.) Simplify $3A + 2B - 3C$ If $A = \frac{2}{3}$ and $C = -\frac{1}{3}$

A) 3
B) -3
C) $-\frac{2}{3}$
D) $-\frac{3}{2}$

2.) If $Y = -1$ and $Z = -2$ what is the value of $5 + 3y - 4z$

A) 10
B) 8
C) -5
D) -16

3.) A Prime Energy drink has 300 mg of caffeine which is 10 times as much caffeine as a young person should ingest. Write an equation to show the amount of caffeine (C) a young person should ingest.

A) $C = 300(10)$
B) $C = \frac{300}{10}$
C) $C = 300 - 10$
D) $C = 300 + 10$

4.) Solve for X: $-1 > -\frac{x}{8}$

A) $-8 > x$
B) $8 > x$
C) $-8 < x$
D) $8 < x$

5.) If X is an integer, which could not be a value for X given: x^2

A) 25
B) 50
C) 81
D) 100

6.) Supplementary angles add up to 180 degrees. Angles J and P are supplementary. Angle J is 65 degrees. How many degrees is angle P?

A) 65
B) 105
C) 115
D) 125

7.) If $c^2 = (a^2)(b^2)$ and $a = 3$ and $b = 4$, what is the value of c?

A) 9
B) 12
C) 13
D) 14

8.) If $c^2 = (a^2)(b^2)$ and $a = 2$ and $b = 7$, what is the value of c?

A) 12
B) 13
C) 14
D) 15

9.) The quotient of 48 and a number X is 6. Solve for X.

A) 5
B) 6
C) 7
D) 8

10.) Joey can do 5 sit ups in 12 seconds. Set up a proportion to show how many sit ups (S) he can do in a minute.

A) $\frac{5}{12} = \frac{s}{60}$
B) $\frac{12}{5} = \frac{s}{60}$
C) $\frac{5}{60} = \frac{12}{60}$
D) $\frac{12}{60} = \frac{s}{5}$

11.) What is the value of $x^4 + x$ if x=-3?

A) -15
B) 81
C) 84
D) 78

12.) Given $\frac{a}{2} = \frac{b}{3} = \frac{c}{6} = 12$, what is the value of a+b+c?

A) 24
B) 36
C) 72
D) 132

13.) 3x + 8 = 29 Solve for X

A) 7
B) 8
C) 9
D) 10

14.) Given X = -1 and Y = 3.5
Solve -4y - 2x

A) -12
B) 12
C) 14
D) -14

15.) What is the domain of the given function

(4,-3), (3,-4), (2,-6), (-1,8)
A) -3, -4, -6, 8
B) 4, 3, 2, 1
C) 4, 3, 2, -1
D) 4, -3, 3, -4

PRACTICE TEST #9

Each Practice Test will have four sections. Each Section will have 15 questions which you will have 10 minutes to answer.

Good Luck!

SECTION 9.1

1.) 96x13 + 13x4

A) 1300
B) 1265
C) 1261
D) 1248

2.) At their end of the season party, the Rowdy Cicadas eat 478 chicken nuggets in the first hour, 136 nuggets in the second hour, and 248 the final hour. One of the Cicadas likes the nuggets so much he takes 26 nuggets home in his backpack! The team ordered 1,000 nuggets. How many nuggets will be left over and sadly thrown out?

A) 96
B) 112
C) 138
D) 144

3.) A cheetah can run a lap around the high school track in 12 seconds. How many laps can he run if he runs at a constant speed for 2 minutes?

A) 24
B) 15
C) 12
D) 10

4.) Potato Salad is normally $6.00 per pound. They made too much at Trader Joe's and put it on sale for 25% off. What is the sale price of the potato salad today?r

A) 4.00$ per pound
B) 4.50$ per pound
C) 4.75$ per pound
D) 5.25$ per pound

5.) What is $\frac{2}{25}$ as a decimal and a percent?

A) $\frac{8}{100}$, 0.08 0.8%
B) $\frac{8}{100}$, 0.8, 8%
C) $\frac{8}{100}$, 0.08, 8%
D) $\frac{8}{100}$, 0.8, 80%

6.) Thomas is bringing his epic homemade salsa for game day with the boys! His secret recipe calls for 3 pounds of garlic for every 27 tomatoes. If he only has 1.5 pounds of garlic on hand, how many tomatoes will he need to pick from his garden?

A) 12
B) 13
C) 14
D) 27

7.) The number of Dallas fans at the Washington game was five times the number of Washington fans. There were 72,000 people at the game. How many Washington fans actually showed up?

A) 60,000
B) 22,000
C) 14,400
D) 12,000

8.) Dog treats are sold in packs of 24 for $13.20 at PetSmart. How much are you paying per dog treat at PetSmart?

A) $0.50
B) $0.55
C) $0.60
D) $0.65

9.) On your trip to Hawaii to go visit your friend who moved there, you bring four pairs of pants, five shirts, and two pairs of sunglasses. How many possible different outfits could you make?

A) 40
B) 32
C) 22
D) 14

10.) Liam skates 20 yards in 15 seconds at his hockey game. If he keeps skating at this rate, how far will he skate in his 3 minute shift?

A) 240
B) 120
C) 80
D) 60

11.) Deer can devastate any unprotected suburban garden by eating 6 plants in a mere 15 seconds! If every plant has an average of 5 leaves, how many leaves will the deer eat in 2 minutes?

A) 900 leaves
B) 180 leaves
C) 240 leaves
D) 90 leaves

12.) A class of 25 students want to buy 10 gallons of slime to dump on their unsuspecting teacher at field day. Each gallon of slime costs $22. If the cost of the slime is split evenly among all the students, how much will each student pay?

A) $40.50
B) $12.60
C) $10.40
D) $8.80

13.) $\dfrac{7^{-2} \times 7^7}{7^2}$

A) 7^3
B) 7^4
C) 7^5
D) 7^6

14.) The high temperature in Yakutsk yesterday was -46 degrees F. If every day for the next six days the temperature increases 2.5 degrees, what will the temperature be?

A) -61
B) -40
C) -31
D) -15

15.) Which inequality is true?

A) $9.81 \times 10^{-5} > 9.81 \times 10^{-2}$
B) $6.73 \times 10^{4} < 6.73 \times 10^{3}$
C) $1.11 \times 10^{8} < 1.11 \times 10^{-9}$
D) $1.93 \times 10^{-3} > 1.93 \times 10^{-4}$

SECTION 9.2

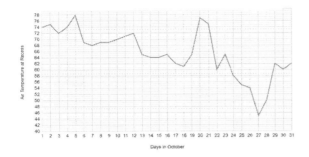

Use the graph above to answer questions 1-4

1.) How much did the temperature decrease from October 1st to the 15th?

A) 12
B) 10
C) 3
D) stayed the same

2.) During which two days was there no change in air temperature?

A) 8th and 9th
B) 1st and 2nd
C) 13th and 14th
D) 15th and 16th

3.) What was the range of the temperature in the month of October?

A) 30
B) 33
C) 29
D) 23

4.) Between which two days was the greatest temperature decrease?

A) 21st and 22nd
B) 26th and 27th
C) 5th and 6th
D) 23rd and 24th

Elon Musk owns several companies that make him the following amounts of money during a given week.(In Millions)

Day	Tesla	SpaceX	Twitter	Open Ai
Mon	$1.7	$0.8	$2.6	$0.8
Tues	2.3	1.2	1.4	0.8
Wed	0.8	1.6	3.6	0.8
Thurs	0.1	0.4	1.4	0.8
Fri	2.2	0.9	1.0	0.8

5.) How much money did Elon make from Open AI that week?

A) $0.8
B) $1.6
C) $2.4
D) $4.0

6.) Which company made Elon the most money on Tuesday?

A) Twitter
B) Space X
C) Tesla
D) Open Ai

7.) What was the mean amount of money made by Twitter Monday through Friday?

A) $10.0
B) $2.0
C) $1.8
D) $1.6

8.) Space X has a test rocket explode and they have to buy a lot of new parts! Space X has to spend $2.5 Million on the parts this week, so how much profit did Space-X actually make?

A) $4.9
B) $4.4
C) $3.6
D) $2.4

Reading a Histogram

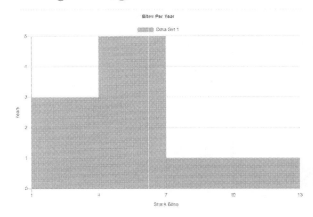

Use the histogram above for questions 9-12 showing how often people are bitten by sharks every year.

9.) What is the mode number of Shark Bites Per Year?

A) 1-4
B) 4-7
C) 7-10
D) 10-13

10.) How many total years are included in the data set?

A) 10
B) 15
C) 20
D) 25

11.) What percent of years did sharks bite 1-4 times?

A) 10%
B) 20%
C) 30%
D) 40%

12.) How many years have there been more than 7 shark bites?

A) 5
B) 4
C) 3
D) 2

13.) Diego catches crabs at the beach, and instead of eating them, attempts to race them! The Blue crab moves 120 inches across the beach. The Red crab moves 5 yards across the beach. The Orange crab then moves 12 feet and 8 inches. List the crabs in order from least to greatest of how far they moved across the beach. Note: 1 yard = 36 inches and 1 Foot = 12 inches.

A) Red, Orange, Blue
B) Orange, Red, Blue
C) Red, Blue, Orange
D) Blue, Orange, Red

14.) Five Neighborhood children are picking tomatoes for Mr. Blake while he is on a two week vacation and can not tend to his garden. He pays them $0.75 per tomato they pick. The children tell Mr. Blake, he owes them $105 total. Assuming the children's math is correct, what is the mean number of tomatoes picked by each child?

A) 140
B) 70
C) 28
D) 14

15.) 5, 6, 7, 8, 5, 6, 7, 8…
What is the 15th term in this sequence?
A) 5
B) 6
C) 7
D) 8

SECTION 9.3

1.) Write an equation for the relationship A is 7 more than half the value of B

A) $A = 2B - 7$
B) $A = \dfrac{B}{2-7}$
C) $A = 2B + 7$
D) $A = \dfrac{B}{2} + 7$

2.) Tessa's Tanning Salon charges $48 per one hour session. Tessa has (S) sessions booked this week. She has to pay a weekly rent of $350 for her business. Write an equation to show how many (D) dollars Tessa will make this week.

A) $D = 350S - 48$
B) $D = 48S + 350$
C) $D = 48S - 350$
D) $D = 350S + 48$

3.) $X + X + X = 48$
$X + X + Y = 26$
What is the value of Y ?

A) 4
B) 6
C) -6
D) -8

4.) Which two ordered pairs that would be in Quadrant II.

A) (-2,2), (-7,7)
B) (-2,-2), (-7,7)
C) (-2,-2), (-7,-7)
D) (2,-2), (7,-7)

5.) Which table shows the number of cookies (C) sold is 4 less than 2 times the number of brownies (B) sold at the Swim Meet.

A)
| C | 4 | 0 | 2 | 6 |
| B | 0 | 2 | 3 | 5 |

B)
| C | 0 | 2 | 3 | 5 |
| B | -4 | 0 | 2 | 6 |

C)
| C | 0 | 5 | 7 | 11 |
| B | -4 | 2 | 3 | 4 |

D)
| C | 4 | 5 | 8 | 11 |
| B | 0 | 2 | 3 | 4 |

6) Troy has 8 fewer cards than Camden. Sarah has three times as many cards as Troy. If Camden has (C) cards, write an expression that shows how many cards Sarah has.

A) 3C-24
B) C-24
C) 3C
D) C-8

7.) Which Scatter plot below does not represent a function?

A)

B)

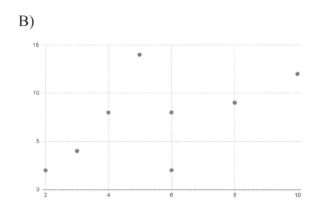

C)

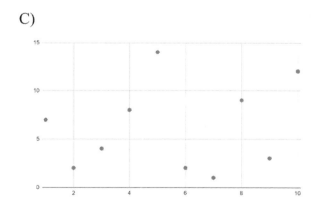

D)

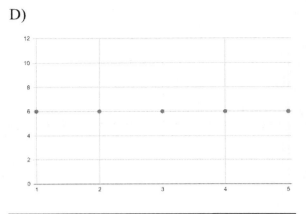

8.) Two premier league players average 16 goals a season. A third player is added to the average, and now the average goals scored per season is 14. How many goals did the third player score that season?

A) 16
B) 14
C) 12
D) 10

9.) The local Community Center charges a flat fee of $300 to rent their space for an evening but gives a discount of $7.50 per person who attends that lives in the community. Write an equation to show the total cost (C) for a group of (P) people to rent the Community Center for an evening.

A) C=7.5P-300
B) C=7.5P+300
C) C=300-7.5P
D) C=-7.5P-300

10.) Which line graph does not show a function?

A)

B)

C)

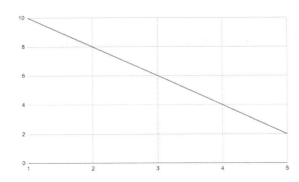
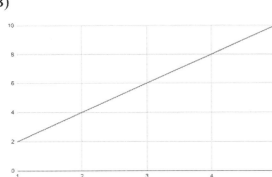
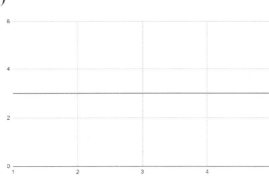

D)

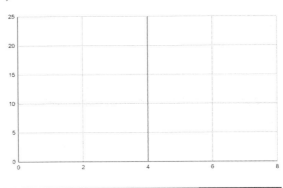

11.) Which ordered pair works for the equation y = 4x + 2

A) (6,1)
B) (1,6)
C) (3,3)
D) (-5,-1)

12.) Angelina's mom sends her with $25 to the local bakery to buy bread. She can buy baguettes (X) for $3.25 and bagels (Y) for $2.50. Write an equation to show how much money she has after buying apples and peaches.

A) M= 3.25x+2.5y+25
B) M= - 3.25x - 2.5y+25
C) M= - 3.25y - 2.5x+25
D) M= 3.25x+2.5y-25

13.) Empress Tzu Tzu has to be taken to the dog groomer, but when her owner takes her to the store, their credit card machine is down, and they are only accepting cash. His owner has $40 cash. The dog groomer charges $5 per dog plus $1.50/minute. How long can Empress Tzu Tzu be groomed for?

A) 20 minutes
B) 23 minutes
C) 25 minutes
D) 27 minutes

14.) If the radius of circle P is 3 inches and the radius of circle Q is 9 inches, given that Area of a Circle is πr2, how much larger is circle Q than circle P?

A) 72π
B) 27π
C) 9π
D) 6π

15.) Birthday Parties at Totally Tubular Skiing & Tubing cost (T) Total Dollars where X is the number of people attending the birthday and each person costs $28. You also have a coupon for $75 off because you got stuck in the chair lift last time. Write an equation for the cost of the party.

A) T=75x+28
B) T=28x+75
C) T=28x-75
D) T=-28x+75

SECTION 9.4

1.) If the absolute value of z is 12, give the two possible integers that z could be equal to

A) 6, 6
B) -6, -2
C) 10, 2
D) 12, -12

2.) If a function R of B= B - R, what is 7 of -4?

A) 3
B) -3
C) -11
D) 11

3.) Carlos can eat $\frac{3}{4}$ of a bag of takis in one minute. Write an expression that shows how many bags of takis he can eat in X minutes.

A) $\frac{4}{3}x$
B) $\frac{3}{4}x$
C) $x+\frac{3}{4}$
D) 3x+4

4.) Write an expression to model -1 is decreased by a number, W.

A) -1-W
B) W+1
C) W-1
D) 1-W

5.) If x= - 6 what is the value of $\frac{x}{x^4}$

A) $\frac{1}{-216}$
B) $\frac{1}{216}$
C) -216
D) 216

6.) If M=HS where M = total miles driven, H represents how many hours you drive, and S represents the speed you drive in the car. If the Ambulance is driving my wife and I to the hospital because she is about to have a baby, how long will it take the ambulance to travel the 15 miles to the hospital if we drive an average of 60 miles an hour?

A) H=15
B) H=0.25
C) H=4
D) H=60

7.) If x = -12 then what is $\frac{x^2}{24}$

A) 12
B) -12
C) 6
D) -6

8.) If $\frac{x}{8}$ = 2, what is the value of $\frac{1}{4x}$?

A) 16
B) 12
C) 8
D) 4

9.) Solve for G: -5G>20

A) G>-4
B) G>4
C) G<-4
D) G<4

10.) What is 3B x 3B?

A) $3B^2$
B) $6B^2$
C) $9B^2$
D) 9B

11.) Rewrite the equation so that y is a function of x. $10y - 8 = 4x - 6y$

A) $y = \frac{x}{4} + \frac{1}{2}$
B) $y = 4x + 8$
C) $y = 4x - 8$
D) $y = \frac{x}{4} + 8$

12.) Simplify the expression: $(-12y)(-5y)$

A) $-60y$
B) $60y^2$
C) $7y$
D) $-7y$

13.) Solve the equation $\frac{2}{3}(9-y)$ If $y=45$

A) -36
B) 36
C) -24
D) 24

14.) The sides of a quadrilateral are x, y, z and 15. If the perimeter of the quadrilateral is 45, write an equation for the perimeter of the quadrilateral in simplest form.

A) $x+y+z=45$
B) $x+y+z+45=15$
C) $x+y+z-15=45$
D) $x+y+z=30$

15.) Write the expression for the quotient of three times a number x and five times a number y.

A) $\frac{3x}{5y}$
B) $\frac{5y}{3x}$
C) $(5y)(3x)$
D) $3x-5y$

220

PRACTICE TEST #10

Each Practice Test will have four sections. Each Section will have 15 questions which you will have 10 minutes to answer.

Good Luck!

SECTION 10.1

1.) $\dfrac{5350}{50}$

A) 106
B) 107
C) 160
D) 170

2.) **If you leave food inside your desk, the classroom might get ants. Twenty two people in the afternoon class get ants in their desks. It costs $60 to have exterminators clean out four desks. How much will it cost to have all the ants exterminated?**

A) $330
B) $660
C) $760
D) $1320

3.) Solve: $\dfrac{42}{\frac{7}{5}}$

A) 3
B) 29.4
C) 30
D) 58.8

4.) Takeshi consistently makes $\dfrac{5}{9}$ shots he takes in basketball games. If he shoots 81 times this season, how many shots did he make?

A) 66
B) 59
C) 55
D) 45

5.) $\dfrac{37.08}{1.2}$

A) 30.9
B) 39
C) 309
D) 390

6.) Your grandpa falls asleep at 8:45 PM and wakes up at 5:25 AM to watch his favorite cable news show. How long was he asleep for?

A) 3 hours and 20 minutes
B) 8 hours and 20 minutes
C) 8 hours and 40 minutes
D) 9 hours and 40 minutes

7.) 8.8 + 0.0098 + 0.909

A) 9.7088
B) 9.7108
C) 9.7188
D) 9.7888

8.) What is the 6th number in this sequence? 3, 9, 27, 81... ???

A) 243
B) 333
C) 666
D) 729

9.) What is the Greatest Common Factor of 72 and 108?

A) 3
B) 9
C) 18
D) 36

10.) There are four blue gatorades and five yellow gatorade bottles. What is the probability of selecting a blue gatorade?

A) 44%
B) 50%
C) 55%
D) 56%

11.) The rectangular shaped floor of the classroom is 150% wider than it is long. If the length of the classroom is 24 feet, what is the perimeter of the classroom?

A) 174 Feet
B) 120 Feet
C) 60 Feet
D) 36 Feet

12.) 0.99 x 0.45

A) 0.4444
B) 0.5555
C) 0.4455
D) 0.4545

13.) Students are playing a math counting game where someone gets out every time they say a number with a 4 in it. The students count to 100. How many people will be out of the game?

A) 19
B) 20
C) 4
D) 10

14.) $-8^2 - 5(-9)$

A) -19
B) 19
C) 109
D) -109

15.) $\dfrac{56.7}{12.6}$

A) 0.045
B) 0.45
C) 4.5
D) 45

SECTION 10.2

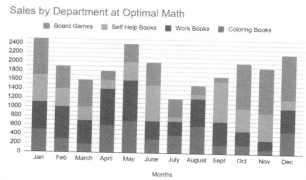

Use the stacked bar chart above for questions 1-6 about monthly sales by department at Optimal Math.

1.) Which sector made the least money in July?

A) Self Help Books
B) Work Books
C) Coloring Books
D) Board Games

2.) What is the mean amount of Work Book Sales in January - March?

A) 500
B) 600
C) 700
D) 800

3.) What is the median amount of Board Game Sales in August - October?

A) 100
B) 200
C) 300
D) 900

4.) Which sector gave Optimal Math the least sales between January and March?

A) Board Games
B) Work Books
C) Self Help Books
D) Coloring Books

5.) Which month did Optimal Math sell the most Self Help Books?

A) May
B) June
C) July
D) August

6.) What was the range of sales revenue from Coloring Books from March - May?

A) 600
B) 500
C) 400
D) 200

7.) Rafael is in a boat when they hit a rock, and the boat gets a hole in it! He quickly begins bailing out the boat while his friends Jack and Theo start rowing them back to shore. Every minute the boat takes on 75 cups of water. Rafael can bail out the boat at a rate of 58 cups per minute. If it will take his friends 30 minutes to row the boat to shore, how much water will be in the boat?

A) 390 cups
B) 430 cups
C) 470 cups
D) 510 cups

8.) Five boys are arm wrestling each other at lunch. Each boy must arm wrestle each other boy once to leave no doubt who is the best arm wrestler! How many arm wrestling contests must there be to determine the winner?
A) 8
B) 10
C) 12
D) 15

9.) Phil, an early adopter of crypto, is doing his taxes and he knows he bought 2 BTC @ $2,500 each, but doesn't know what he paid for the rest. He sells all 5 of his BTC for a total of $200,000! Coinbase tells him that he has taxable profits of $150,000. So what price did he buy his other 3 BTC each at? (Assume the remaining three were bought at the same price)
A) $50,000 per BTC
B) $45,000 per BTC
C) $25,000 per BTC
D) $15,000 per BTC

10.) Listed below is a chart of how many chicken sandwiches Chick Fil A sold this week. The manager was not at work on Wednesday, but does know they sold 2,800 chicken sandwiches this week. (All Chick Fil A's are closed on Sunday.) How many chicken sandwiches were sold on Wednesday?

M	T	W	TH	F	S
400	340	???	390	350	750

A) 570
B) 530
C) 470
D) 430

11.) A local Taco Shop did a taste test on the street of people's favorite type of Taco they sell:

Favorite Taco	Number of People
Carne Asada	12
Pollo Al Carbon	4
Barbacoa	6
Al Pastor	8
Chorizo	10
Pescado	4
Chicharron	6

What percent of students chose Barbacoa?
A) 6%
B) 12%
C) 18 %
D) 24%

12.) At Marine Corp boot camp the trainees do 250 pull ups in two days. If they continue at this pace, how many pull ups will they do their first week of boot camp?
A) 875
B) 750
C) 625
D) 1,000

13.) Students were asked to track how much time they spent doing the following activities outside of school.

Activity	% of Time Spent
Sleeping	71%
Playing Soccer	2%
Studying for IAAT	25%
TV/Video Games	1%
Watering the Garden	1%

If students have about 16 hours outside of school every day, how many hours do they spend studying for the IAAT?

A) 8 hours
B) 6 hours
C) 4 hours
D) 2 hours

14.) My wife folds the laundry four times faster than I do. We started folding laundry at the same time and have folded a combined total of 30 pieces of laundry. How many pieces of laundry did I personally fold?

A) 0
B) 5
C) 6
D) 7.5

15.) Ximena studied for her IAAT for 2 hours on Monday. She studied three times as long on Tuesday as she did on Monday. On Wednesday she studied 5 hours less than she did on Tuesday. On Thursday she studied for half as long as she had on Wednesday. How many total hours did she study for her IAAT that week?

A) 8.5 hours
B) 9 hours
C) 9.5 hours
D) 10 hours

SECTION 10.3

1.) Fill in the chart for the function y=5x-5

x	0	1	2	3
y				

A) -5, 0, 5, 10
B) -5, 5, 10, 15
C) -5, 5, 25, 50
D) 5, 10, 25, 50

2.) What is the next term in the sequence: 9, 11, 14, 18, _____

A) 22
B) 23
C) 24
D) 25

3.) What is the rule?

In	-1	0	3	N
Out	4	3	0	????

A) y=-x-3
B) y=x-3
C) y=x+3
D) y=-x+3

4.) What is the X intercept of the line y = -2x - 4?

A) -2
B) -4
C) -6
D) -8

5.) Write an equation for one third the number of band students (B) is twelve more than twice the number of students in strings (S).

A) $B=\frac{3}{2}S-12$

B) $B=\frac{2}{3}S+12$

C) $\frac{B}{3}=2S+12$

D) $B=\frac{2}{3}S-12$

6.) $\frac{1}{8} > Y > \frac{1}{10}$
Give a possible value for Y

A) 0.11
B) 0.22
C) 0.33
D) 0.44

7.) Make a table that shows the number of Hot Dogs sold (yH) is three less than twice the number of Cookies sold (xC).

A)

C	0	2	3	4
H	-3	-1	3	5

B)

C	0	2	3	4
H	-3	1	3	5

C)

C	0	2	3	4
H	-3	-1	3	6

D)

C	0	3	4	4
H	3	3	12	6

8.) What is the slope of a line
10x - 2y = -10

A) -10
B) 10
C) 5
D) -5

x	3	5	7	9	12
y	12	20	28	36	???

9.) If X is 12, what is Y?

A) 48
B) 56
C) 63
D) 72

10.) When X=-3, what is the value for y in the equation $y=\frac{4}{3}x - 2$

A) -12
B) -4
C) -6
D) -8

11.) Alan (A) scores eight more goals than Maggie (M). Write an equation for how many goals Maggie scores.

A) $M=\frac{A}{8}$
B) M=A-8
C) M=A+8
D) M=8A

12.) Given the function, what is the value for B when A=16?

A		4	8	12	16
B		3	6	9	???

A) 15
B) 14
C) 12
D) 10

13.) Write the equation that matches the verbal statement: the number of car riders (C) is two more than one sixth of the number of walkers (W).

A) $C=\frac{1}{6}W+2$
B) C=6W+2
C) $\frac{1}{6}C=W+2$
D) $C+2=\frac{1}{6}W$

14.) If $y=\frac{1}{3}x -9$ and x=18, what is the value of y?

A) 9
B) 3
C) 0
D) -3

15.) In their run away win last night, the number of points scored by the Purple Cobras is three less than double the number of points scored by the Pink Pandas. If the Pink Pandas scored 28 points, how many points did the Purple Cobras score?

A) 50
B) 53
C) 56
D) 59

SECTION 10.4

1.) Solve for Y: $\frac{3}{4}(y-8) = 9$

A) 20
B) 18
C) 15
D) 12

2.) The Optimal Shoe Company makes Profit (P)= Revenue (R) - Fixed Costs (C). Optimal's investors will give them a nice bonus if they can produce 1.3×10^5 in profit this month. If Optimal Shoes has fixed costs of 9.2×10^4 this month, how much do they need to sell so their investors give them that sweet bonus?

A) 7.9×10^3
B) 7.9×10^4
C) 3.8×10^3
D) 3.8×10^4

3.) Solve for Y: $-15y+18=-27$

A) y=-3
B) y=3
C) y=24
D) y=-24

4.) If Angle A is a right angle and the measure of Angle A is 15X, what is the value of X?

A) 90
B) 30
C) 6
D) 5

5.) The Raging Kangaroos average 7 goals every 4 games. Set up a proportion to show how many goals he scores in a 9 game season.

A) $\frac{7}{4} = \frac{4}{x}$
B) $\frac{7}{4} = \frac{x}{9}$
C) $\frac{7}{4} = \frac{9}{x}$
D) $\frac{4}{7} = \frac{7}{9}$

6.) Solve: $2z + 7 < 13$

A) 3
B) 4
C) 5
D) 6

7.) The students get three times as rambunctious every day they have to stay inside for indoor recess. If they have had indoor recess four days in a row, in terms of R, how more rambunctious are they than normal?

A) $R4^3$
B) $4R^3$
C) $3R^4$
D) 12R

8.) How many factors does 36 have?

A) 4
B) 6
C) 8
D) 10

9.) If 3x=12 and x - y= -12, what is the value of y?

A) 16
B) -8
C) 8
D) -16

10.) Four more than the product of five and a number (Z) is negative twenty six. Solve for Z

A) -4
B) 4
C) 6
D) -6

11.) The volume of a cone can be found using the formula $V=\frac{1}{3}H\pi r^2$
What is the volume of the cone if the radius is 5 inches and the height is 27 inches.

A) 175π
B) 195π
C) 225π
D) 315π

12.) Z^3=216, what does Z^2=

A) 36
B) 25
C) 16
D) 9

13.) The sum of all the interior angles in all triangles is 180 degrees. In triangle CDE, Angle C measures 115, Angle D is 35, how many degrees is Angle E?

A) 20
B) 25
C) 30
D) 35

14.) If $\frac{x}{4}$ - 4y = 20 and y=-2, what is x equal to?

A) 60
B) 48
C) -24
D) -36

15.) You are on a long car trip with your family. Your younger cousin constantly asks "are we there yet?" You explain to him that you had traveled x miles, 25 miles ago. You tell him you will reach your destination in 35 more miles. How long is the total trip in terms of x?

A) x-10
B) x-60
C) x+10
D) x+60

PRACTICE TEST ANSWERS

Test 1

Section 1.1
1.) C
2.) B
3.) B
4.) A
5.) C
6.) B
7.) D
8.) A
9.) B
10.) A
11.) D
12.) B
13.) B
14.) C
15.) A

Section 1.2
1.) A
2.) D
3.) A
4.) B
5.) B
6.) A
7.) B
8.) C
9.) D
10.) B
11.) D
12.) B
13.) C
14.) C
15.) C

Section 1.3
1.) C
2.) A
3.) A
4.) C
5.) D
6.) A
7.) C
8.) A
9.) B
10.) D
11.) B
12.) C
13.) B
14.) A
15.) C

Section 1.4
1.) A
2.) A
3.) A
4.) B
5.) B
6.) D
7.) C
8.) B
9.) B
10.) B
11.) B
12.) D
13.) D
14.) D
15.) A

Test 2
Section 2.1
1.) A
2.) D
3.) A
4.) C
5.) C
6.) B
7.) C
8.) A
9.) C
10.) B
11.) C
12.) A
13.) B
14.) D
15.) D

Section 2.2
1.) C
2.) D
3.) C
4.) A
5.) C
6.) D
7.) D
8.) C
9.) A
10.) D
11.) D
12.) D
13.) B
14.) A
15.) B

Section 2.3
1.) B
2.) C
3.) C
4.) A
5.) B
6.) C
7.) A
8.) C
9.) C
10.) A
11.) D
12.) B
13.) B
14.) A
15.) A

Section 2.4
1.) C
2.) C
3.) B
4.) B
5.) B
6.) A
7.) A
8.) C
9.) B
10.) C
11.) A
12.) A
13.) D
14.) A
15.) C

Test 3
Section 3.1
1.) C
2.) B
3.) B
4.) C
5.) C
6.) A
7.) A
8.) C
9.) B
10.) C
11.) B
12.) A
13.) B
14.) D
15.) B

Section 3.2
1.) A
2.) C
3.) D
4.) B
5.) A
6.) C
7.) A
8.) D
9.) C
10.) C
11.) A
12.) C
13.) B
14.) C
15.) D

Section 3.3
1.) C
2.) C
3.) C
4.) A
5.) B
6.) B
7.) C
8.) B
9.) B
10.) A
11.) B
12.) C
13.) C
14.) B
15.) A

Section 3.4
1.) B
2.) A
3.) A
4.) C
5.) A
6.) B
7.) C
8.) D
9.) A
10.) B
11.) B
12.) C
13.) C
14.) B
15.) A

Test 4
Section 4.1
1.) A
2.) B
3.) D
4.) D
5.) D
6.) A
7.) B
8.) D
9.) D
10.) B
11.) C
12.) A
13.) B
14.) A
15.) C

Section 4.2
1.) C
2.) B
3.) A
4.) D
5.) A
6.) A
7.) B
8.) D
9.) A
10.) C
11.) Decreases
12.) B
13.) B
14.) C
15.) D

Section 4.3
1.) D
2.) D
3.) A
4.) C
5.) A
6.) D
7.) A
8.) D
9.) C
10.) A
11.) A
12.) B
13.) C
14.) C
15.) A

Section 4.4
1.) D
2.) C
3.) A
4.) C
5.) B
6.) B
7.) C
8.) C
9.) A
10.) C
11.) B
12.) B
13.) C
14.) B
15.) B

Test 5
Section 5.1
1.) B
2.) B
3.) A
4.) B
5.) C
6.) D
7.) A
8.) C
9.) C
10.) C
11.) C
12.) A
13.) B
14.) D
15.) C

Section 5.2
1.) B
2.) A
3.) A
4.) D
5.) C
6.) C
7.) C
8.) B
9.) C
10.) D
11.) C
12.) D
13.) B
14.) A
15.) B

Section 5.3
1.) B
2.) A
3.) C
4.) D
5.) A
6.) B
7.) C
8.) C
9.) C
10.) B
11.) D
12.) C
13.) A
14.) D
15.) B

Section 5.4
1.) D
2.) B
3.) A
4.) B
5.) B
6.) D
7.) D
8.) A
9.) C
10.) B
11.) D
12.) D
13.) B
14.) D
15.) B

Test 6
Section 6.1
1.) C
2.) B
3.) B
4.) D
5.) D
6.) B
7.) C
8.) C
9.) A
10.) C
11.) A
12.) A
13.) D
14.) A
15.) A

Section 6.2
1.) B
2.) A
3.) B
4.) B
5.) A
6.) B
7.) C
8.) C
9.) D
10.) C
11.) B
12.) C
13.) D
14.) A
15.) D

Section 6.3
1.) B
2.) C
3.) A
4.) B
5.) D
6.) A
7.) B
8.) D
9.) A
10.) C
11.) D
12.) B
13.) D
14.) B
15.) B

Section 6.4
1.) D
2.) B
3.) C
4.) B
5.) A
6.) C
7.) A
8.) A
9.) B
10.) D
11.) C
12.) C
13.) A
14.) D
15.) D

Test 7
Section 7.1
1.) D
2.) B
3.) D
4.) C
5.) D
6.) D
7.) C
8.) A
9.) B
10.) A
11.) B
12.) B
13.) A
14.) A
15.) C

Section 7.2
1.) C
2.) A
3.) D
4.) B
5.) A
6.) B
7.) C
8.) C
9.) A
10.) D
11.) B
12.) C
13.) A
14.) C
15.) D

Section 7.3
1.) C
2.) A
3.) A
4.) D
5.) D
6.) C
7.) B
8.) B
9.) C
10.) B
11.) D
12.) A
13.) B
14.) D
15.) D

Section 7.4
1.) C
2.) B
3.) D
4.) A
5.) B
6.) D
7.) A
8.) C
9.) A
10.) D
11.) D
12.) D
13.) A
14.) B
15.) C

Test 8
Section 8.1
1.) B
2.) A
3.) D
4.) B
5.) C
6.) A
7.) C
8.) A
9.) B
10.) B
11.) D
12.) C
13.) A
14.) C
15.) B

Section 8.2
1.) D
2.) A
3.) B
4.) A
5.) B
6.) C
7.) C
8.) B
9.) A
10.) B
11.) A
12.) C
13.) B
14.) D
15.) A

Section 8.3
1.) B
2.) A
3.) A
4.) D
5.) A
6.) D
7.) B
8.) D
9.) C
10.) A
11.) B
12.) A
13.) A
14.) B
15.) D

Section 8.4
1.) D
2.) A
3.) B
4.) D
5.) B
6.) C
7.) B
8.) C
9.) D
10.) A
11.) D
12.) D
13.) A
14.) A
15.) C

Test 9
Section 9.1
1.) A
2.) B
3.) D
4.) B
5.) C
6.) C
7.) D
8.) B
9.) A
10.) A
11.) C
12.) D
13.) A
14.) C
15.) D

Section 9.2
1.) B
2.) A
3.) B
4.) A
5.) D
6.) C
7.) B
8.) D
9.) B
10.) A
11.) C
12.) D
13.) A
14.) C
15.) C

Section 9.3
1.) D
2.) C
3.) C
4.) A
5.) A
6.) A
7.) B
8.) D
9.) C
10.) D
11.) B
12.) B
13.) B
14.) A
15.) C

Section 9.4
1.) D
2.) C
3.) B
4.) A
5.) A
6.) B
7.) C
8.) D
9.) C
10.) C
11.) A
12.) B
13.) C
14.) D
15.) A

Test 10
Section 10.1
1.) B
2.) A
3.) C
4.) D
5.) A
6.) C
7.) C
8.) D
9.) D
10.) A
11.) B
12.) C
13.) A
14.) C
15.) C

Section 10.2
1.) A
2.) B
3.) B
4.) D
5.) B
6.) B
7.) D
8.) B
9.) D
10.) A
11.) B
12.) A
13.) C
14.) C
15.) C

Section 10.3
1.) A
2.) B
3.) D
4.) A
5.) C
6.) A
7.) B
8.) C
9.) A
10.) C
11.) B
12.) C
13.) A
14.) D
15.) B

Section 10.4
1.) A
2.) D
3.) B
4.) C
5.) B
6.) A
7.) C
8.) C
9.) A
10.) D
11.) C
12.) A
13.) C
14.) B
15.) D

Made in the USA
Middletown, DE
22 November 2024

65237656R00139